Antibiotics
A Multidisciplinary Approach

Antibiotics
A Multidisciplinary Approach

Giancarlo Lancini

Marion Merrell Dow Research Institute (MMDRI)
Lepetit Research Center
Gerenzano (Varese), Italy

Francesco Parenti

Marion Merrell Dow Europe AG
Horgen, Switzerland

and

Gian Gualberto Gallo

Formerly of Marion Merrell Dow Research Institute (MMDRI)
Lepetit Research Center
Gerenzano (Varese), Italy

Springer Science+Business Media, LLC

Library of Congress Cataloging-in-Publication Data

Lancini, Giancarlo.
Antibiotics : a multidisciplinary approach / Giancarlo Lancini, Francesco Parenti, and Gian Gualberto Gallo.
p. cm.
Includes bibliographical references and index.
ISBN 978-1-4757-9202-7 ISBN 978-1-4757-9200-3 (eBook)
DOI 10.1007/978-1-4757-9200-3
1. Antibiotics. I. Parenti, Francesco. II. Gallo, Gian Gualberto. III. Title.
[DNLM: 1. Antibiotics--pharmacology. 2. Antibiotics--metabolism. 3. Structure-Activity Relationship. 4. Drug Resistance, Microbial. QV 350 L249a 1995]
RM267.L33 1995
615'.329--dc20
DNLM/DLC
for Library of Congress 95-11625
CIP

Publication history

First edition: *Biochimica e Biologie degli Antibiotici*, G. C. Lancini and F. Parenti, ISEDI, Milan, Italy (1977)

First edition–English translation: *Antibiotics–An Integrated View*, G. C. Lancini and F. Parenti, Springer-Verlag, New York (1982)

First edition–Russian translation: *Antibiotiki*, G. C. Lancini and F. Parenti, MIR, Moscow, USSR (1985)

Second edition: *Biochimica e Biologia delgi Antibiotici*, G. C. Lancini, F. Parenti, and G. G. Gallo, Salerno, Italy (1993)

ISBN 978-1-4757-9202-7

Originally published by Plenum Press, New York in 1995
Softcover reprint of the hardcover 1st edition 1995

10 9 8 7 6 5 4 3 2 1

Preface

Fifty years after the introduction of penicillin into therapeutic use, antibiotics are among the most widely used drugs, not only in the treatment of human ailments but also in veterinary practice, in agriculture, and in animal husbandry. Furthermore, they are essential tools for research in genetics, microbiology, and molecular biology.

Production of antibiotics involves a diverse group of professionals: the microbiologist, the fermentation technologist, the bioengineer, and the chemist. Research and development of a new antibiotic requires the collaboration among experts in even more numerous disciplines: genetics, molecular biology, industrial microbiology, medical microbiology, chemistry of natural products, and analytical chemistry.

In teaching undergraduate and postgraduate courses, it became apparent that a book did not exist presenting the facts and the basic concepts of this broad and varied subject. During many years of experience in an industrial laboratory dedicated to the research of new antibiotics, we realized that our young colleagues, specialized in one discipline, often had a poor knowledge of the general aspects of the research in which they were involved and of the problems brought about by the development and the use of antibiotics.

This book was born eighteen years ago to fulfill, at least in part, this need. Some years later, updated translations in English and in Russian were published, and, to celebrate the fiftieth anniversary of the introduction of penicillin into medical use, a new Italian edition, completely revised in content and format, was published in 1993. The present version, which is essentially based on the last Italian edition, has been greatly modified from the first English edition, so that it is practically a new book.

We would like to believe that reading this book may also interest medical doctors, who, either in laboratories or clinics, use antibiotics and are curious to know more about the mental and practical processes that lead from the isolation of a microorganism from a soil sample to a tablet or a vial ready to use. The understanding of the broader "biology of antibiotics" may also be of help in the proper use of these drugs.

Many research colleagues in both the industrial and the academic worlds have helped conceive and realize this book. We thank them and acknowledge their contribution and the greater merit for its potential usefulness. We are particularly indebted to Dr. Mark Fisher, St. George's Hospital, London, and Stefano Donadio, of our Laboratories, for their criticisms and suggestions, and to Ms. Suzanne Ress, who patiently revised the English style.

As for the first edition, we wish to dedicate this book to Piero Sensi, discoverer of rifamycins, teacher, and friend.

G. C. Lancini, F. Parenti, and G. G. Gallo

Contents

Chapter 1

The Antibiotics

An Overview

The aim of this chapter is to provide the reader with basic definitions and general information on antibiotics in order to facilitate an understanding of the following chapters, where various aspects of the biology and chemistry of antibiotics are described in a more specific way.

Thus, this chapter can be considered an expanded summary of the book; it give the picture of the whole before presenting its component parts.

1.1. Definition

Antibiotics are low-molecular-weight microbial metabolites that at low concentrations inhibit the growth of other microorganisms. A low-molecular-weight substance is a molecule with a defined chemical structure having a relative mass of at most a few thousand. We do not include among the antibiotics those enzymes, such as lysozyme, and other complex protein molecules, such as colicins, that also have antibacterial properties.

If the definition were to be adhered to rigorously, the only substances to be considered antibiotics would be the natural products of microorganisms. However, currently products obtained by chemical modification of natural antibiotics or of other products of microbial metabolism are called semisynthetic antibiotics. Even the requirement that antibiotics be "microbial metabolites" is no longer strictly applied. For example, one often sees expressions such as "antibiotic products of plants," but we feel that this usage of the term is inaccurate.

With the expression "inhibition of the growth of other microorganisms" by an antibiotic, we mean either temporary or permanent inhibition of the ability of the microorganism to reproduce and, consequently, inhibition of growth of the bacterial population rather than of an individual cell. When inhibition is permanent, antibiotic activity is termed *bactericidal* or, more generally, *cidal* (e.g., fungicidal). If inhibition is lost when an antibiotic is removed from its medium, the antibiotic is said to have a *bacteriostatic* (fungistatic) or *static action*.

The specifier "at low concentration" is added to the definition, as even essential and normal cell components can cause damage at excessive concentrations. For example, glycine, one of the constituents of every protein, has a strong bactericidal effect on some bacteria if it is present in the culture medium in high concentrations. Similarly, ethanol or butanol, the fermentation products of some microorganisms, cannot be considered antibiotics because they display antibacterial activity only at high concentrations. With the term "low concentration" we generally mean values well below 1 mg/ml.

1.2. Chemical Nature

To date, about 10,000 antibiotics have been isolated and described and the chemical structures of the majority have been determined. For the remaining ones, sufficient knowledge of their activities and their main physicochemical properties is available to allow us to identify their structures. It is clear that chemically, antibiotics are a very heterogeneous group.

1. As a group, antibiotics include substances of molecular weight from 150 to 5000.
2. Their molecules may contain only carbon and hydrogen, or, more commonly, carbon, hydrogen, oxygen, and nitrogen; others also contain sulfur, phosphorus, or halogen atoms.

3. Almost all the organic chemical functional groups are represented (hydroxyl, carboxyl, carbonyl, nitrogen functions, etc.) as are all the organic structures (aliphatic chains, alicyclic chains, aromatic rings, heterocycles, carbohydrates, polypeptides, etc.).

From these heterogeneous aspects it is clear that antibiotics are not a homogeneous category of chemical substances, unlike steroids or proteins. A statement such as "I am allergic to antibiotics" is nonsense, because allergy is a phenomenon connected with the presence of specific chemical groups in the allergen and there are no chemical groups common to all antibiotics, even in the smaller group of antibiotics used in therapy.

The only property that all antibiotics have in common is that they are *organic solids*. Organic is obvious from their definition as products of microbial metabolism. It is less obvious why liquid antibiotics should be almost unknown. Molecules that are rather large or have several polar groups are solids at room temperature. Usually antibiotics have several polar groups, which are involved in their interaction with bacterial macromolecules resulting in inhibition of bacterial growth. We must therefore accept this as the reason why even the smaller antibiotic molecules are solid substances.

The relationships between the chemical structures of antibiotics and their activities are discussed in Chapter 5.

1.3. Producing Microorganisms

This large variety of molecules is produced by an array of widely diverse microorganisms, but the taxonomic distribution of the strains that produce them is not uniform. More than 50% of the antibiotics described are in fact produced by members of only one bacterial order, Actinomycetales, and particularly by one genus of this order, *Streptomyces*. The eubacteria rarely produce antibiotics, except the sporogenic bacilli that produce a particular class of antibiotics, the peptides, and species of the genus *Pseudomonas*. Only two genera of fungi, the *Aspergillus* and the *Penicillium*, produce a relatively high number of antibiotics; in many other genera of Fungi Imperfecti less than ten antibiotics per genus are known.

The production of antibiotics is not rigorously species specific: the same antibiotic can be produced by organisms belonging to different species or genera or even orders. And the reverse is also true; that is, strains classified taxonomically as members of the same species can produce different antibiotics. However, as a general rule, the more distant the organisms are on the taxonomic scale, the less probable it is that they will produce the same antibiotic.

The relationship between the producing microorganisms and the antibiotics produced is discussed in Chapter 9.

1.4. Biosynthesis

In contrast to the great variety of chemical structures and producing strains, the biological reactions involved in the synthesis of antibiotics can be grouped into a few fundamental biosynthetic pathways. It is important to recognize that these pathways are simple variations of the biosynthetic pathways of normal cellular metabolism, and it is surprising that small changes in these pathways can give rise to such diverse substances.

A common feature of the reactions involved in the biosynthesis of antibiotics, and more generally of *secondary metabolites*, is that the substrate–enzyme specificity frequently appears to be less rigid than in reactions involved in the biosynthesis of *primary metabolites*. In the biosynthesis of secondary metabolites, a given enzyme may catalyze the same reaction in the presence of several slightly different substrates. On the other hand, a single metabolite may be transformed into different products by reactions catalyzed by different enzymes. This partial lack of specificity results in the synthesis of products with a common basic structure but differing, for example, in degree of oxidation or methylation. For this reason, antibiotics are often produced in *families*, i.e., the same strain makes two or more antibiotics that are structurally related.

On the basis of biosynthetic pathways, antibiotics can be grouped as follows:

1. Analogues of primary metabolites (analogues of amino acids, nucleosides, coenzymes, etc.). These are small molecules that are biosynthesized in a manner similar to that of the primary metabolites and that often resemble them structurally.
2. Antibiotics derived by polymerization. These include:
   a. *Peptide* antibiotics and their derivatives, produced from the condensation of some amino acids to form a polypeptide chain that can be modified by further reactions (it is important to note that in most cases the condensation of the amino acids *does not* take place by the classic protein synthesis mechanism).
   b. Antibiotics derived from *acetate* and *propionate* units. There is a wide variety of chemical structures, but all are

derived from reactions that follow the biosynthetic pathways of fatty acids.

c. *Terpenoid* antibiotics derived from isoprene synthesis (these are produced by fungi and only exceptionally by actinomycetes).

d. *Aminoglycoside* antibiotics, derived by condensation of a few sugar molecules, frequently amino sugars, and a cyclic amino alcohol (amino cyclitol).

In addition, the biosynthetic pathway described for some antibiotics is not easily classifiable, and there are known antibiotics that are derived from condensation of subunits originating from more than one of the pathways mentioned above.

The principal biosynthetic pathways are described in Chapter 6.

1.5. Activity and Resistance

Antibiotics are frequently grouped according to their *spectrum of activity,* that is, according to the classes of microorganisms they inhibit. There are, therefore, *antiviral, antibacterial, antifungal,* and *antiprotozoal antibiotics,* according to the main groups of sensitive microorganisms. One speaks also of *antitumor antibiotics,* products of microbial origin that inhibit the growth of cancer cells. The use of the term *antibiotic* for these products is justified by the fact that these products were frequently isolated on the basis of their antibacterial activities.

The sensitivity of different bacteria to antibiotics depends largely on the structure of their cell walls, as it determines the ability of the antibiotic to penetrate the bacterial cell. Therefore, the antibacterial antibiotics can be divided according to activity against gram-positive or gram-negative bacteria or mycobacteria. There are many more antibiotics effective against gram-positive bacteria, which are easily permeable: these antibiotics are said to have a *narrow spectrum of activity* . If they are active against both gram-positive and gram-negative bacteria, they are said to have a *broad spectrum of activity.*

In a given population individual microorganisms may be present that are not inhibited at antibiotic concentrations that inhibit the majority of the cells. These individuals are called *mutants.* Their frequency in an otherwise sensitive population is different for different species and different antibiotics and ranges between one mutant in every 10^7 to over 10^{10} sensitive cells.

If a microbial population that contains resistant mutants is exposed to inhibitory concentrations of an antibiotic, growth of the sensitive cells is blocked while the resistant mutants continue to multiply so that eventually the whole population will be made up of resistant cells. Thus, the antibiotic can effectively act to select resistant mutants and promote the development of a resistant population. This phenomenon, which can be easily observed in the laboratory, occurs also in the environment.

In a population of resistant mutants the formation of sensitive cells may occur with a high frequency by back mutation. Often the rate of growth of sensitive cells is higher than that of resistant mutants and, thus, in the long term, in the absence of the antibiotic, the population reverts back to sensitivity.

When a microorganism is resistant to two or more antibiotics (usually structurally related) it is said to be cross-resistant to them.

Some bacteria have been found to be able to transfer the property of resistance to given antibiotics to other bacteria of the same species and even of different species. This transfer is part of the more general phenomenon of exchange of genetic material from one cell to another that takes place by different mechanisms in different groups of microorganisms. The result is that in the presence of a resistant and "infectious" cell a microbial population can become resistant to an antibiotic without going through the normal selection process. This is known as *transferable resistance*.

The main methods for determining the activity of antibiotics are described in Chapter 2 and the phenomenon of resistance is discussed in Chapter 4.

1.6. Mechanism of Action

Antibiotics block the growth of sensitive microorganisms by inhibiting the action of a molecule, usually a macromolecule, such as an enzyme or a nucleic acid, essential for cell multiplication. At the molecular level this means that the antibiotic molecule is able to bind to a specific site on the target macromolecule, forming a *molecular complex*, which is no longer able to accomplish its original function.

To determine an antibiotic's mechanism of action, one has to identify the target macromolecule and its function. It is usually easier to identify the function that is blocked than the particular macromolecule involved, and for this reason we speak of antibiotics that inhibit cell wall, protein, or RNA synthesis, DNA replication, or membrane function, depending on what appears to be the primary effect of the antibiotic.

Some antibiotics are *antimetabolites*, acting as competitive inhibitors. These are structurally similar to normal metabolites, such as amino acids or coenzymes, and bind to the enzyme for which the metabolite is substrate or cofactor, thus inactivating it.

The antibiotic's selectivity of action is usually related to its mechanism of action. This topic is examined in Chapter 3.

1.7. Chemotherapy

Chemotherapy, the drug treatment of infectious diseases, is based on the ability of antibiotics (and of some other chemical substances) to inhibit the multiplication of the infecting microorganism without an intolerable toxic effect on the cells or the metabolic functions of the human organism. This inhibition makes it easier for the body's defenses to overcome infection. We have already stated that about 10,000 antibiotics have been isolated, and, by modification of the most interesting among them, tens of thousands of compounds have been synthesized chemically and tested for antimicrobial activity. However, only a few have been found to possess the characteristics necessary for clinical use, which can be classified as follows:

1. *Activity against one or more pathogenic microorganisms*. It is desirable that a broad-spectrum or a narrow-spectrum antibiotic have low propensity to select resistance and it is preferable, though not essential, that it also have bactericidal effect.
2. *Good absorption and distribution*. To be effective, an antibiotic must be absorbed, reach the site of the infection and remain there at inhibitory concentrations for a sufficient time. It must be eliminated from the body within a reasonable time to avoid accumulation and potential toxic consequences.
3. *Lack of toxicity*. The antibiotic must be without any untolerable toxicity to the host at therapeutic doses. Serious adverse reactions are tolerable only if the antibiotic is to be used in diseases that are extremely severe or potentially lethal.

The concepts and the principal methods used to determine these properties for new antibiotics are described in Chapter 7. The clinical use of antibiotics is discussed in Chapter 8.

1.8. Chemical Modifications

In relatively complex molecules such as antibiotics, some structural components or chemical groups are directly involved in the formation of a complex with the macromolecular target of their action, while other structural components or groups are not directly involved in this and, therefore, can be modified chemically without substantially modifying the antibiotic's *intrinsic activity*. By means of these alterations one can modify some physicochemical characteristics of the molecule, especially those involved in water or lipid solubility, which, in turn, modify the metabolic and pharmacokinetic properties. These same properties may also affect the spectrum of activity, as they influence the antibiotic's ability to penetrate into the cell. Through chemical modification, products that are active against resistant mutants can be obtained.

These are the main reasons for the large-scale effort to chemically modify natural antibiotics, described in Chapter 5. This work has been of great importance, during the years following the discovery of the first antibiotics, in providing new products for therapeutic use.

1.9. Principal Classes of Antibiotics

Various schemes for classification of antibiotics have been proposed, none of which have been universally adopted. Currently, those natural or semisynthetic antibiotics that have a common basic chemical structure are grouped into one "class" and named after the member first discovered or after a principal chemical property. This empirical classification is very useful in practice, as the components of one class usually share many biological properties. A detailed description of the properties of the different classes of antibiotics is given in Chapter 5.

1.9.1. β-Lactam Antibiotics (Penicillins and Cephalosporins)

The penicillins were the first antibiotics to be used in therapy and are still considered the drugs of first choice for treatment of many infections. The penicillins and the more recently developed cephalosporins form the group of β-*lactam antibiotics*, so called because of the presence in their molecule of a four-atom cyclic amide, chemically called β-lactam. They were first isolated from fungi of the genera *Penicillium* and *Cephalosporium*. Later, actinomycetes of the genera *Streptomyces* and *Nocardia* and also some gram-negative bacteria, have been shown to produce β-lactam antibiotics. From a biosynthetic aspect, the β-lactam antibiotics may be considered to be derived from the polymerization of amino acids.

The β-lactam antibiotics inhibit the synthesis of peptidoglycan (see Section 3.3.3), a basic component of the bacterial cell wall, causing irreversible damage. They are therefore bactericidal antibiotics. They are not active against fungi, whose cell wall does not contain peptidoglycan, or against mycoplasma, which lack a cell wall.

The spectrum of action of the early penicillins, such as the widely known *penicillin G*, was limited to gram-positive bacteria and some gram-negative cocci and, furthermore, a rapid diffusion of resistant strains was observed. Successively, by means of chemical modifications, derivatives have been obtained that are effective to varying degrees against almost all gram-negative bacteria. The same results have been obtained with the cephalosporin derivatives starting from the weakly active original *cephalosporin C*. Other objectives have been obtained following the wide research carried out in this field, i.e.:

1. The preparation of penicillins and cephalosporins active when given orally
2. The preparation of derivatives somewhat less sensitive to the β-lactamases, which are drug-inactivating enzymes produced by some bacterial species or by some resistant bacterial strains
3. The isolation of new β-lactam antibiotics produced by streptomycetes and other bacteria, referred to as *nonclassic β-lactams*, which, together with their semisynthetic derivatives, have an enlarged spectrum of activity against the less sensitive bacteria

With certain exceptions, the toxicity of the penicillins and cephalosporins is very low.The major problem in the use of these antibiotics is the appearance of hypersensitization phenomena and allergy, sometime with very severe manifestations.

1.9.2. Tetracyclines

This family of antibiotics is characterized by its very broad spectrum of action and by its great therapeutic effectiveness. It originally included only *chlortetracycline*, *oxytetracycline*, and *tetracycline*, the last being the one most widely used clinically.

The tetracyclines are products of different strains of *Streptomyces*. They are biosynthesized by cyclization of a chain obtained by condensation of acetate and malonate units. The chemical structure consists of four rings condensed linearly, and this is the basis for their name.

They act by preventing ribosomal protein synthesis. The effect is reversible and, therefore, they are bacteriostatic agents. Their activity spectrum is particularly broad, including gram-positive and gram-negative bacteria, rickettsiae, chlamydiae, and some protozoa. Because of their physicochemical properties (they are insoluble at neutral pH), the natural tetracyclines can only be given orally. A great deal of study on semisynthetic products resulted in the preparation of injectable derivatives and derivatives with prolonged action.

Only limited success has been achieved in the search for derivatives with activity against resistant bacteria, which, nowadays, are common.

1.9.3. Aminoglycoside Antibiotics (Aminocyclitols)

These form a large class of substances produced by members of the genera *Streptomyces*, *Micromonospora*, and *Bacillus*. They are characterized chemically by the presence of a cyclic amino alcohol to which some amino sugars are bound. Both the amino alcohol and the amino sugars are derived biosynthetically from glucose.

The aminoglycoside antibiotics inhibit protein synthesis irreversibly by interacting with ribosomes, with a bactericidal effect. They are particularly active against gram-negative bacteria. *Streptomycin*, the first aminoglycoside known, was discovered by means of a systematic research program aimed at isolating an antibiotic effective against gram-negative organisms. Streptomycin was also the first antibiotic effective against tuberculosis.

Intensive research programs carried out in the last decades, aimed at isolating derivatives active against the numerous resistant strains, have allowed us to isolate or to synthesize, among others, *kanamycin*, *gentamicin*, *tobramycin*, *amikacin*, and *netilmicin*, which are active against bacteria insensitive to streptomycin.

Because of the presence of numerous hydroxyl groups in their molecule, all the aminoglycosides are very water-soluble, which explains their inability to be adsorbed orally.

Their major adverse effects include nephro- and ototoxicity.

1.9.4. Macrolides

The chemical structure of these antibiotics is characterized by a ring consisting of no fewer than 12 carbon atoms and closed by a lactone group. They are typical products of *Streptomyces*. They appear to be biosynthesized by condensation of a number of acetate and propionate units.

The macrolides can be subdivided into two homogeneous classes:

1. *Antibacterial macrolides*, characterized by the presence of lactone rings of 14 or 16 carbon atoms, with at least two sugar molecules. They reversibly inhibit protein synthesis by interacting with the

ribosomes with bacteriostatic action. Their activity spectrum is restricted to gram-positive bacteria and mycoplasma. The typical representative of this class is *erythromycin*. Also included are *oleandomycin*, *leucomycins*, and *spiramycin*, and some newer derivatives that have been synthesized in order to improve oral absorption are also components of this class.

2. *Antifungal and antiprotozoal macrolides*, characterized by lactone rings of about 30 atoms, with hydroxyl substituents and including a series of conjugated double bonds (from 4 to 7). For this last characteristic, they are also called *polyenes* (tetraenes, pentaenes, etc.). They are active only when administered intravenously. They induce distortions in the cell membranes by interfering with sterols. Therefore, they are not active against bacteria, which do not contain sterols in their cytoplasmic membrane, but only against fungi and some protozoa. Their toxicity makes it necessary to limit their use to the most serious cases. The best known representative of this family is *amphotericin B*, a heptaene.

1.9.5. Ansamycins

This is a family of antibiotics, introduced in therapeutic use at the end of the 1960s, which present a typical structure: an aliphatic chain that connects two opposite points of an aromatic ring, like a handle of a basket or ansa (hence the name). They are produced by strains of several genera of the order Actinomycetales. Their biosynthesis resembles that of the macrolide antibiotics, i.e., condensation of a number of acetate and propionate units.

The ansamycins can be divided into two groups, the *naphthalenes* and the *benzenes*, according to the type of aromatic ring present. The more important naphthalene ansamycins are antibacterial and selectively inhibit the enzyme RNA polymerase. The benzene ansamycins are less selective in their action, and have been studied as possible antitumor agents. Interestingly, *maytansine*, which belongs to this class, is a plant product.

Among the naphthalene ansamycins are the rifamycins, which are very active against gram-positive bacteria and mycobacteria. While the natural rifamycins are not used in therapy, some semisynthetic rifamycins have chemotherapeutical roles, such as *rifamycin SV*, which is used against infections of the bile tract, and *rifampicin* (in the United States, *rifampin*), which is active orally and has a broad spectrum of action, being particularly active against *Mycobacterium tuberculosis* and the staphylococci.

The frequency of resistant mutants to rifampicin is rather high, and it varies from species to species.

1.9.6. Peptide Antibiotics

The chemical structure of these antibiotics consists of a chain of amino acids, often closed into a ring. They are produced by a great variety of microorganisms by biosynthetic pathways different from protein synthesis. The first antibiotics isolated were *gramicidin* and *bacitracin,* which are only of historical interest, since they are too toxic to be given systemically. Both of them are active against gram-positive bacteria; the first one interferes with the membrane functions and the second one is an inhibitor of the peptidoglycan synthesis.

Among the peptides used systemically are the *polymyxins,* which are very effective against gram-negative bacteria, by affecting their membrane functions.

Newer members of this class are *daptomycin* and *ramoplanin,* both inhibitors of the cell wall synthesis, which are presently under development.

1.9.7. Glycopeptide Antibiotics (Dalbaheptides)

These are linear heptapeptides, having at least five of the amino acid residues represented by benzene rings, which are connected together to form diphenyl and triphenyl ether groups. These rings bear various substituents, like hydroxyls, chlorine atoms, and sugars. Their mechanism of action involves the formation of a complex with the terminal D-alanyl-D-alanyl moiety of the intermediates in bacterial cell wall synthesis. Resistant mutants are very rare. The proposed name dalbaheptides derives from the two described characteristics (D-alanyl-D-alanyl-binding heptapeptide). The important members of this class are: *vancomycin,* produced by *Amycolatopsis orientalis,* and *teicoplanin,* a mixture of five very similar substances produced by *Actinoplanes teichomyceticus,* which are active against gram-positive bacteria, in particular staphylococci resistant to the β-lactams.

1.9.8. Antitumor Antibiotics

These belong to different chemical classes but all act on DNA replication by different mechanisms. Of historical importance are the *actinomycins,* constituted by a phenoxazine ring bearing as substituents two identical peptide cycles. They intercalate between the DNA bases with the aromatic groups, affecting DNA functionality. In the eukaryotic cells they mainly interfere with RNA synthesis.

The clinically important members of the class are the anthracyclines (*daunorubicin* and *doxorubicin*), which have a tetracyclic structure bearing a sugar substituent. They cause breakages in the DNA filaments by interfering with the enzyme topoisomerase. *Mitomycin* C has a tricyclic structure and acts by forming covalent bonds bridging between two DNA filaments. *Bleomycin*, a mixture of nine peptide components produced by *Streptomyces verticillus*, causes breakages in the DNA filaments by a quasi-enzymatic reaction.

1.9.9. Miscellaneous Antibiotics

Some antibiotics used in therapy cannot be classified in any of the families so far described. Some of these are:

Chloramphenicol, originally isolated from *Streptomyces venezuelae*, has since been produced synthetically. It is one of the few natural compounds that contain a nitro group. It inhibits protein synthesis and is bacteriostatic. It is very effective in the cure of infections with gram-negative bacteria, especially against salmonellae. It is active orally.

Lincomycin resembles the antibacterial macrolides in its mechanism of action (inhibition of protein synthesis) and in its spectrum of activity (limited to gram-positive bacteria), and in fact it demonstrates a partial cross-resistance to erythromycin. Chemically, it is completely different, being made up of a modified amino acid condensed with a complex amino sugar. It is particularly active against some anaerobic bacteria. *Clindamycin*, one of its semisynthetic derivatives, has similar properties.

Novobiocin, active against gram-positive bacteria and *Proteus*, interferes with DNA by inhibiting one of the subunits of the enzyme gyrase.

Fusidic acid, produced by a fungus, has a steroid-type structure and is active against gram-positive microorganisms, by inhibiting one of the elongation factors in protein synthesis. It is orally active. The principal limitation in its usefulness is the high frequency of resistant mutants.

Fosfomycin, initially obtained from *Streptomyces griseus* and now produced by chemical synthesis, inhibits the peptidoglycan synthesis and is bactericidal. It possesses a rather wide spectrum of action, but shows a high frequency of resistant mutants.

Griseofulvin is an antimycotic drug that can be given systemically. It is produced by a fungus and has an aromatic structure, derived biosynthetically from condensation of acetate and malonate units. It is an inhibitor of cytoskeleton formation.

Mupirocin is a relatively recent antibiotic, purified from *Pseudomonas fluorescens* extract, whose activity has been known since the beginning of this century. It inhibits protein synthesis by interfering with the enzyme lysyl-tRNA synthetase. It has a broad spectrum of action, but is not active systemically, only topically.

Chapter 2

The Activity of Antibiotics

The activity of an antibiotic reflects its ability to inhibit microbial growth. While the concept of *antibiotic activity* appears simple and straightforward, the *quantitative expression* of antibiotic activity is complex and depends on the assay method and the conditions under which the test microorganism is grown.

This chapter gives a brief account of the commonly used techniques of *quantitative determination* of antibiotic activity. However, the aim is not to describe the techniques of determination, but to present interpretative and working principles of these techniques, in order to give the reader a key to understanding the significance and the limitations of the quantitative expression of the activity. The interpretative and working principles of the measurement of antibiotic *concentration* by methods based on its *activity* are also discussed.

2.1. Definition

The *activity* of an antibiotic is defined and measured in terms of its ability to inhibit microbial growth (bacteria, fungi, and protozoa). While the concept of growth is familiar when applied to macroscopic organisms (irreversible increase in volume as a function of time), it must be redefined when applied to microscopic organisms. In this case, two different levels of growth can be distinguished: population growth and growth of a single cell. By population growth we mean the increase with time of the number of microorganisms, i.e., the increase in the density of the microbial population. By cell growth we mean the synthesis of cellular material needed for one cell to give rise to two daughter cells. Obviously, the growth of a microbial population is the result of cell growth.

The activity of an antibiotic is normally defined as its ability to inhibit the growth of a microbial population. Activity is measured either by directly counting the number of microorganisms per unit volume or by determining some parameters of the culture that are related to population density, such as the property of scattering light.

The growth inhibition of a bacterial population by an antibiotic can be either a reversible or an irreversible phenomenon. In the former case when the antibiotic is removed, the majority of cells start again to reproduce and the action of the antibiotic is said to be *bacteriostatic*; in the latter case, none or only a small fraction of cells remain viable and the action of the antibiotic is said to be *bactericidal*.

The techniques here described, referring to antibacterial activity, can be generally adapted to measure the activity against yeasts, fungi, and sometimes protozoa.

2.2. Determination of the Activity

2.2.1. Inhibitory Activity

One method of quantifying the activity of an antibiotic is to determine the minimal concentration needed to completely inhibit the growth of a given bacterial strain. This is called the *minimal inhibitory concentration* (MIC).

2.2.1.1. Determination of the Minimal Inhibitory Concentration in Liquid Medium

According to the traditional method, the MIC is determined as follows:

1. A series of test tubes is prepared, all containing the same volume of medium inoculated with the test bacterium (the inoculum may

vary from 10^3 to 10^6 cells per milliliter). Today, microwells containing 0.1 ml of medium are ordinarily used.

2. Decreasing concentrations of antibiotic are added to the tubes. Usually a stepwise dilution by a factor of 2 is used (i.e., if the concentration of antibiotic in the first tube is 128 μg/ml, in the second tube it will be 64 μg/ml, and in the third 32 μg/ml, and so on). One tube is left without antibiotic, to serve as a positive control for bacterial growth.
3. The cultures are incubated at a temperature optimal for the test bacterium and for a period of time sufficient for the growth of at least 10–15 generations (usually overnight).
4. The tubes are inspected visually to determine where bacteria have grown, as indicated by turbidity (in fact, turbidity of the culture medium is indicative of the presence of a large number of cells, at least 10^7/ml). The tubes in which the antibiotic is present in concentrations sufficient to inhibit bacterial growth remain clear (Figure 2.1). In experimental terms the MIC is the concentration of antibiotic present in the "last" clear tube, i.e., in the tube having the lowest antibiotic concentration in which growth is not observed.

Obviously, because of the way it is determined (stepwise twofold dilutions of the antibiotic and visual determination of growth), the MIC is not a very accurate value. In fact, displacement by only one tube in the determination of growth causes a twofold variation in the MIC. Nevertheless, the MIC is a very useful parameter both for study of the

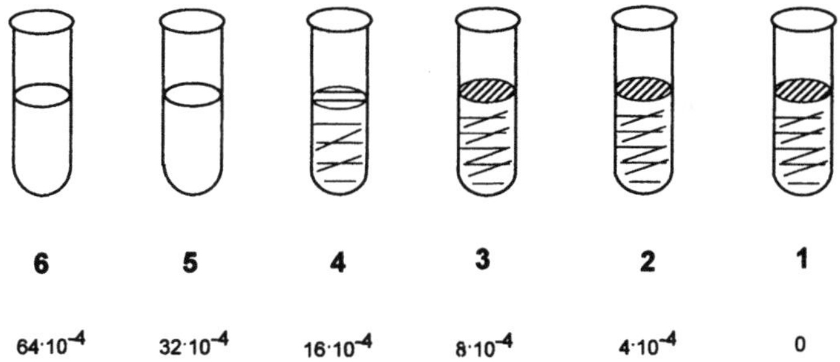

Figure 2.1. Determination of the MIC of rifampin against *Staphylococcus aureus* in liquid medium. Each tube contains 1 ml of culture medium inoculated with 10^4 bacteria/ml. The values below each tube indicate the concentrations of rifampin in μg/ml (tube 1 contains no antibiotic and serves as the positive control for growth). The MIC is the concentration present in tube 5 (0.0032 μg/ml).

antibiotic biology and for its clinical use. However, it has to be kept in mind that the MIC value depends on the experimental conditions and that, within a given species, different strains may be more or less sensitive to the antibiotic and therefore may have different MIC values.

As an example, the MICs determined in liquid medium for some commonly used antibiotics against some laboratory standard strains are reported in Table 2.1.

2.2.1.2. Determination of the Minimal Inhibitory Concentration in Solid Medium

In concept, this is similar to the determination of the MIC in liquid medium. A series of stepwise dilutions of the antibiotic are prepared in a suitable culture medium containing agar and these are distributed in petri dishes. The surface of the medium is then seeded with a culture of the test strain (e.g., 1 μl of a suspension containing 10^9 cells/ml). After a suitable incubation period, the MIC is determined as the lowest concentration at which the total absence of growth, or the formation of few isolated colonies, is observed. One advantage of this method over the liquid medium one is that in this case a single plate can be seeded in different areas with different strains of bacteria, enabling the MIC for several bacteria to be determined in a single operation (Figure 2.2).

2.2.1.3. Cumulative Curve

As mentioned above, different strains of the same species can display a different sensitivity versus the same antibiotic. Thus, in order to evaluate an antibiotic's therapeutic potential it is important to determine its activity against numerous clinical isolates for each species. Usually, the MICs against each strain are determined and the results obtained are expressed as follows:

1. *Activity range*, expressed by both the lower and the higher MICs observed
2. *MIC_{50}*, i.e., the concentration that inhibits 50% of the strains tested
3. *MIC_{90}*, i.e., the concentration that inhibits 90% of the strains tested

The results can also be expressed by a diagram, where the percentage of strains inhibited by each concentration is plotted against the antibiotic concentration. The shape of the cumulative curve indicates the variability of the species's sensitivity against the antibiotic (see Figure 2.3).

Table 2.1. MICs (μg/ml) of Some Representative Antibiotics against Representative Laboratory Strains[a]

Microorganism	Penicillin G[b]	Ampicillin	Ceftazidime[b]	Erythromycin	Tetracycline	Chloramphenicol	Gentamicin	Rifampin	Teicoplanin
Gram-positive									
S. aureus Tour L165	0.06	0.06	8	0.13	0.13	4	0.13	0.016	0.13
S. epidermidis ATCC 12228	128	1	8	0.06	4	1	0.13	0.008	0.13
S. pyogenes C203 SKF 13400	0.03	0.016	0.25	0.016	0.13	0.5	8	0.06	0.06
S. pneumoniae UC 41	0.06	0.03	0.5	0.03	0.06	1	1	0.06	0.06
E. faecalis ATCC 7080	2	0.5	128	0.06	0.13	2	8	0.5	0.13
Gram-negative									
E. coli SKF 12140	64	2	0.5	64	1	1	0.25	8	128
K. pneumoniae 1801	—	—	—	64	2	0.5	0.13	16	128
P. vulgaris ATCC 881	128	2	0.13	128	8	1	0.13	8	128
P. aeruginosa ATCC 10145	128	—	1	128	16	8	0.13	8	128
N. gonorrhoeae L997	0.06	0.03	0.03	0.25	0.06	0.13	2	0.13	32
H. influenzae ATCC 19418	8	0.5	0.25	2	0.25	0.13	1	0.13	64

[a]Determinations carried out in microtiters (100μl) in Iso-Sensitest medium with inocula of about 10^4 CFU/ml.
[b]Inoculum 10^5 CFU/ml.

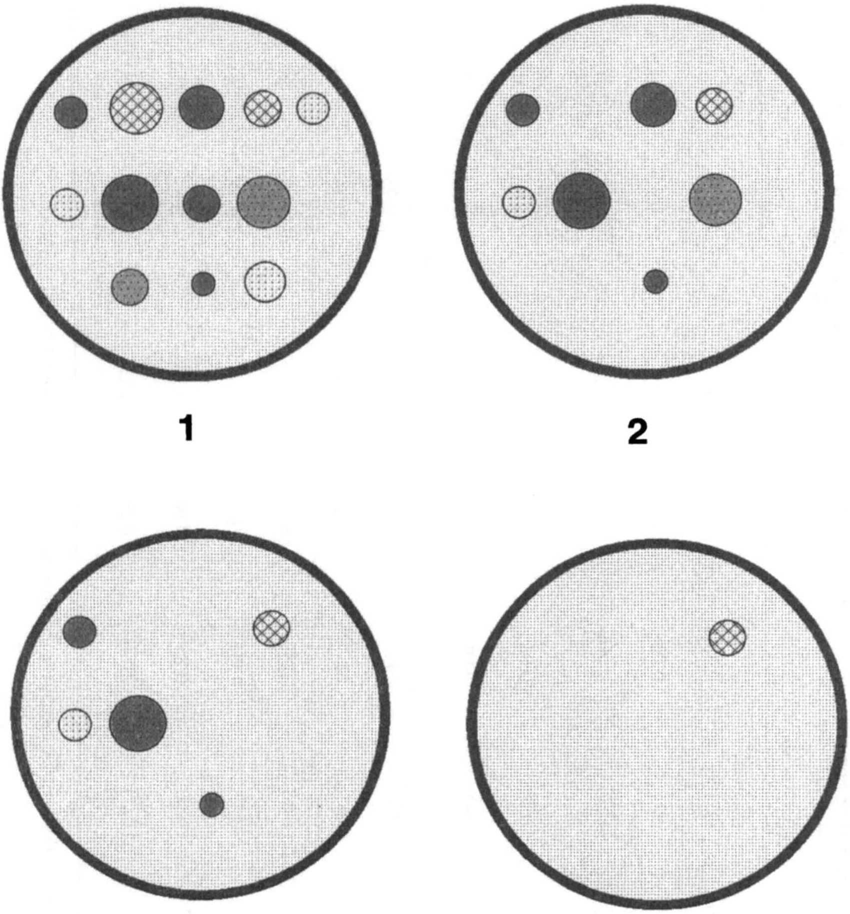

1 2

3 4

Figure 2.2. Simultaneous determination of the MIC for rifampin against 12 species of bacteria in solid medium. Plate 1 is the positive control and contains no antibiotic. In plates 2, 3, and 4 there are increasing concentrations of antibiotic, 0.02, 0.04, and 0.08 μg/ml. One can easily see that different bacteria have different susceptibilities toward the antibiotic. With a sufficiently large number of dilutions, the MICs for all 12 of the bacterial species can be determined at the same time.

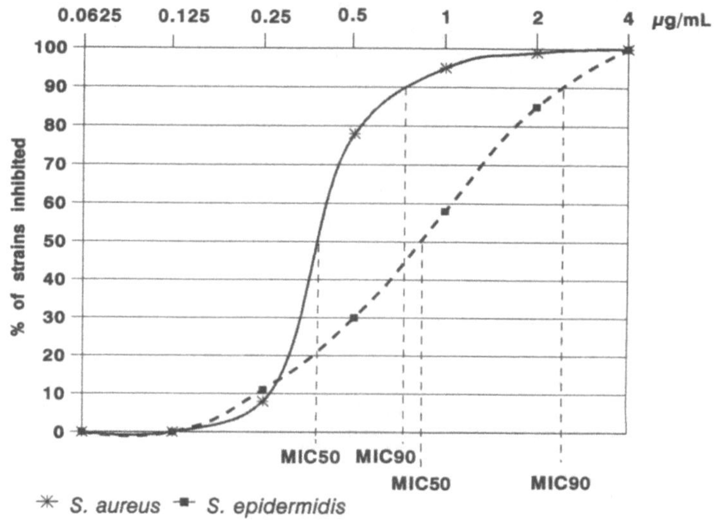

Figure 2.3. Cumulative inhibition curve by teicoplanin for clinical isolates of *S. aureus* (43 strains) and of *S. epidermidis* (57 strains). The antibiotic concentrations are reported on the abscissa and the percentage of the strains inhibited by each concentration is reported on the ordinate.

2.2.2. Determination of Bactericidal Activity

The activity of an antibiotic is defined as *bacteriostatic* when the inhibition of the activity is reversible for a wide concentration range; it is defined as *bactericidal* when the inhibition is irreversible at concentrations slightly higher than the MIC.

2.2.2.1. Minimal Bactericidal Concentration

The *minimal bactericidal concentration* (MBC) is determined as follows:

1. All the operations described for the determination of the MIC in liquid medium are carried out by using an inoculum of at least 10^5 cells.
2. After incubation, an aliquot of medium is taken from each tube in which no growth can be seen and, after appropriate dilution, is used to seed plates containing a suitable agar medium.

3. The plates are incubated for 48 h and the growth of colonies is observed. The number of bacterial colonies is assumed to correspond to that of viable cells. The MBC is the concentration of antibiotic present in the last tube (the one at the lowest antibiotic concentration) from which no colonies grow on the petri plates. In practice, the MBC is arbitrarily defined as the concentration at which a 1000-fold reduction in colony forming units (CFU) is observed with respect to the original inoculum (survival of 0.1%).

2.2.2.2. Killing Curves

One significant parameter of the bactericidal effect is the rate at which it takes place at different antibiotic concentrations. This can be evaluated as follows: a culture containing a large number of bacterial cells (at least 5×10^5/ml) is prepared, a known quantity of antibiotic added, and the culture incubated. At given time intervals, portions are taken and, after appropriate dilution, are plated on agar medium. After 48 h of incubation the colonies grown on the plates are counted and, taking into account the dilution, the number of viable cells is obtained (each viable cell is assumed to give rise to one colony). The operation can be repeated with different concentrations of antibiotic and with different strains. The results are usually presented on a diagram (Figure 2.4) plotting the logarithm of the number of cells able to form colonies (CFU) against time. These curves, called *cumulative killing curves*, give an indication of the efficacy of the bactericidal action in a more realistic way than the MBC does. In fact, they represent the effect of the different concentrations of antibiotic that can be compared to those present in biological fluids after administration for clinical treatment.

2.2.3. Determination of the Susceptibility of Bacteria toward Antibiotics

To carry out a rational antibiotic therapy, it is always useful and often essential to know the antibiotics to which the pathogen causing the infection is susceptible and those to which it is resistant. The determination of the MICs of the various available antibiotics toward the strain isolated from the patient yields the most reliable result. However, the classical MIC methods are not amenable to large numbers of determinations, and simpler and cheaper methods have been developed. Among them, the most common is the *diffusion method* (also known as

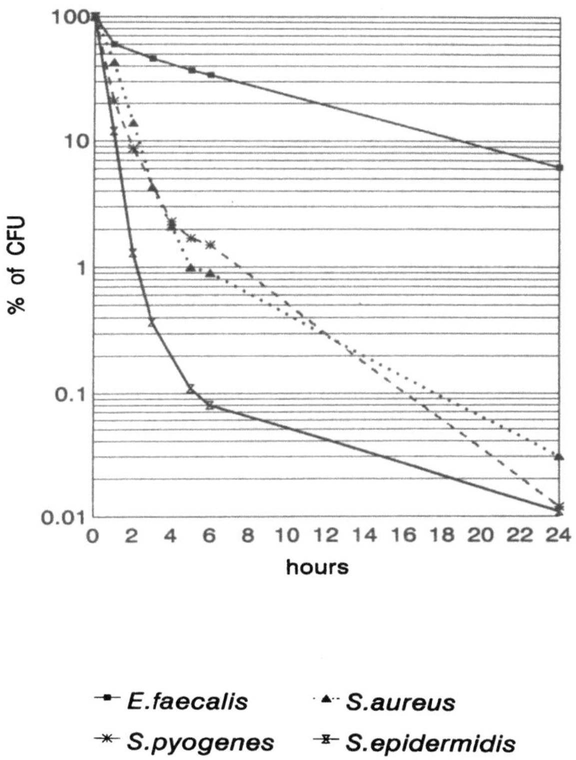

Figure 2.4. Killing curves of teicoplanin against some bacterial strains. Inoculum: about 10^6 cells/ml; concentrations: *E. faecalis* 4 μg/ml, *S. aureus* 2 μg/ml, *S. pyogenes* 0.5 μg/ml, *S. epidermidis* 8 μg/ml.

disk diffusion test or *agar diffusion test*). Another method used in some countries, particularly in the United Kingdom, is the agar incorporation or *critical concentration method*. These methods are here briefly described. However, most laboratories nowadays employ automated methods, which are outlined in Section 2.4.

2.2.3.1. Diffusion Method

In this method filter paper disks soaked with a standard solution of the antibiotic to be assayed are placed on the surface of an agar medium seeded with a dilute and uniform suspension of the bacterium to be tested. The antibiotic diffuses from the disk into the agar, generating a decreasing concentration gradient. The plate is incubated for 16–20 h and the microorganism grows, forming a turbid surface of confluent or semiconfluent colonies. In the zone around the disk where the concentration of the antibiotic is higher than the MIC, the bacterium cannot grow and a transparent inhibition halo is visible. The diameter of this halo relates to the susceptibility of the microorganism to the antibiotic.

The inhibition diameters are measured and compared with the reference ones (breakpoints), whose determination is described in Section 2.2.3.3. The microorganism is generally classified as *sensitive, intermediate,* or *resistant,* according to whether the inhibition halo is larger or equal, slightly smaller, or definitely smaller than the reference one, respectively. In order to be used efficiently, this method needs an accurate standardization. Most frequently applied is the one proposed by Kirby-Bauer.

A variation of this method consists of incubating, at the same time, the strain under examination and a standard strain of the same species, and of using as reference the halo produced by the latter.

2.2.3.2. Method of the Critical Concentration

The principle of this method is analogous to that described for the determination of the MIC in solid medium (Section 2.2.1.2). In practice, the susceptibility of a bacterial strain toward an antibiotic is determined by observing its growth or absence of growth on a predetermined concentration of antibiotic included in an agar plate. Unlike the previous method, this system is of the type "all or nothing." In fact, the strains that form colonies are judged *resistant* and those that do not are judged *sensitive,* without any indication of the degree of resistance or susceptibility, and without any possibility of attributing the category *intermediate.* In order to avoid this inconvenience, the test can be carried out by adding a plate containing a higher concentration of antibiotic. In this way, one can obtain an indication of the level of resistance and classify as intermediate the strains that grow on the plate at lower concentration but not on the plate at higher concentration.

2.2.3.3. Dispersion Diagram and Selection of the Breakpoints

When using the methods for the determination of the susceptibility of bacteria toward antibiotics, it is essential to establish the concentration that discriminates between the categories resistant and sensitive. This concentration depends on a series of parameters that are essentially pharmacological and pharmacokinetic. Among them, the most important (see Chapter 8) are blood levels, elimination rate, and serum protein binding.

Once the discriminating concentration has been established, applying it to the critical concentration method is straightforward and relatively simple. It is more complex to find the discriminating concentration using the diffusion method, because it requires the conversion of the diameter of inhibition halos to concentration data. In order to establish the critical inhibition diameter corresponding to the discriminating concentration (*breakpoint*), the MICs are determined on a large number of strains (from 100 to 200) representing the various pathogenic species. On the same strains the inhibition diameters are also determined with a disk containing a given quantity of antibiotic. Then, one draws a dispersion diagram (*scattergram*) by plotting the logarithms in base 2 of the MIC values against the diameter of the halos. By the mean square method, one can calculate the linear correlation between the diameter and the logarithm of the MICs. On the basis of this correlation the breakpoint is obtained, i.e., the diameter corresponding to the discriminating MIC. The dispersion of the measured values can also be calculated.

2.2.3.4. The Antibiogram

The term *antibiogram* indicates a set of information on susceptibility or resistance of a microorganism toward a panel of antibiotics. Usually the antibiogram is determined by one of the methods described above. In the case of the diffusion method, one can place on the plate seeded with the test strain the standard disks of the different antibiotics (see Figure 2.5). Remember that the dimension of the halo does not depend only on the activity of the antibiotic but, in a relevant way, on its diffusion rate as well. It would be a mistake to automatically consider the antibiotics that yield larger halos more active. In the case of the critical concentration method one can lay, with a multiple seeder, a large number of test strains on a single plate containing the standard concentration of the antibiotic.

A greater accuracy can be obtained when the antibiograms are de-

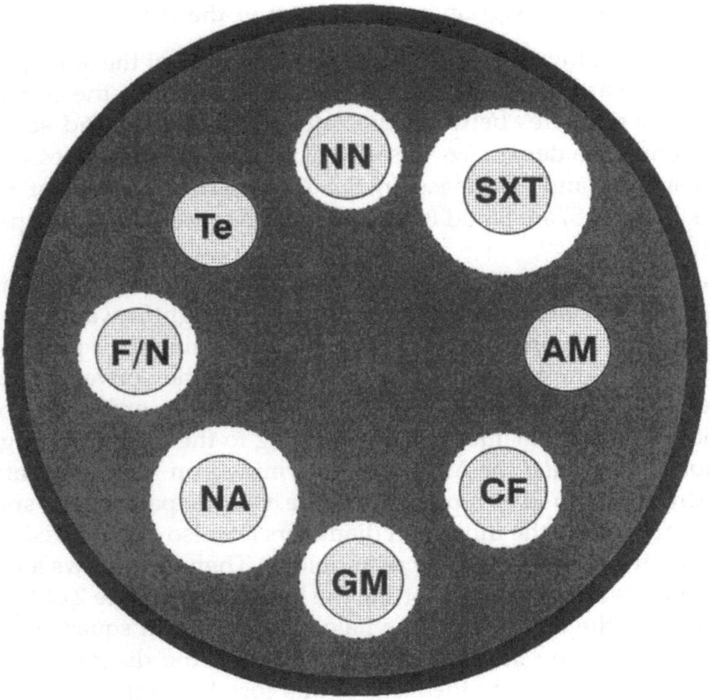

Figure 2.5. Antibiogram. The microorganism is *Escherichia coli*, which can be seen to be sensitive to cephalotin (CF), gentamicin (GM), nalidixic acid (NA), nitrofurantadin (F/N), tobramycin (NN), and co-trimoxazole (SXT). It is resistant to tetracycline (Te) and ampicillin (AM).

termined in liquid medium. Presently, miniaturized systems are frequently used by means of partially or totally automated apparatuses (see Section 2.4).

The choice of the panel of antibiotics and the clinical significance of the antibiogram are discussed in Chapter 8.

2.2.4. Interaction of Antibiotics

The term *interaction* or *interference* refers to the effect that an antimicrobial agent has on the inhibitory action of a second antimicrobial agent when both are added simultaneously (in *combination*) to a microbial culture.

Synergism is assumed when the antimicrobial effect of the combination is greater than the sum of the effects of each component alone;

additivity (or addition) is assumed when the effect of the combination is equal the sum of the effects of the single components. *Antagonism* is assumed when one component decreases the antimicrobial activity of the other. *Indifference* refers to a situation in which the presence of one component (e.g., inactive against the test microorganism) does not alter the activity of the other.

Although these four situations are conceptually clear and distinct, the underlying causes (examined in Chapter 8) and the parameters describing them are complex and often not easily understandable. The quantitative determination of the interactions is not simple and the results depend on the method used and the nature of the parameter determined (MIC or MBC, or variation in growth rate or killing rate, etc.). A method commonly used is that of determining the MICs of combinations of two antibiotics at different concentrations and of expressing these interactions graphically by the use of the *isobologram*. In this diagram the concentration of antibiotic A present in each inhibitory end point is plotted against the concentration of antibiotic B that gives the same end point. Each point of the obtained curve represents the minimal concentrations of A and B that are, *together*, sufficient to inhibit bacterial growth. The shape of the curve indicates what type of interaction is taking place (see Figure 2.6).

The coordinates of an isobologram represent concentrations of antibiotics and not their effect. Therefore, the concept of synergism can be redefined in terms of concentration instead of effect. For this reason, the concentrations are expressed as fractions of the MIC, *fractional inhibition concentration* (FIC), instead of quantity of product per volume. Those combinations for which the sum of the FICs are lower than unity are defined as synergic.

A less accurate but simpler method to detect interaction consists of placing two paper strips each soaked in one antibiotic, arranged at 90°, on the surface of an agar plate seeded with the test microorganism. An increase of the inhibition zone at the intersection of the two strips indicates synergism, a decrease indicates antagonism (Figure 2.7).

2.2.5. Factors Affecting the Determination of Antibiotic Activity

The *in vitro* activity of an antibiotic is affected by the experimental conditions under which it is determined. Some of the relevant factors are: the composition of the medium, the density of the bacterial inoculum, and the total number of bacterial cells in the inoculum. In addition, there are some specific factors that affect activity when determined in solid medium.

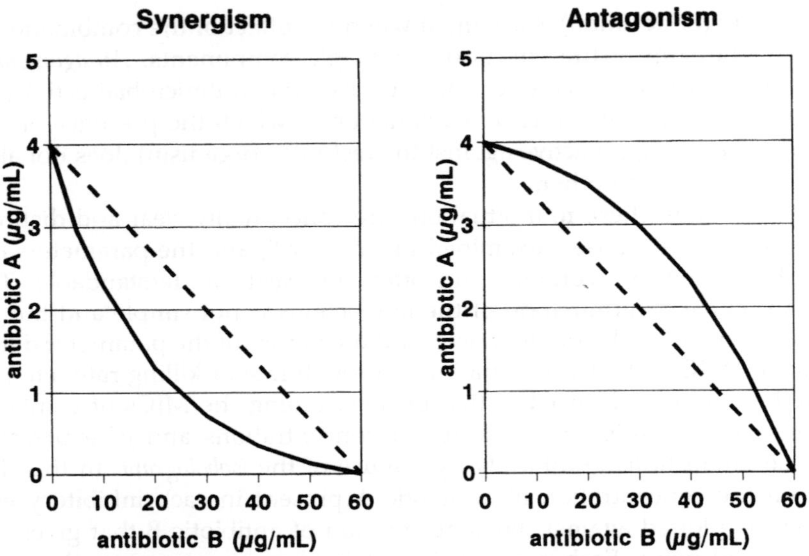

Figure 2.6. Isobolograms indicating synergism or antagonism between antibiotics A and B. The points on the right of the solid curve represent the concentrations of antibiotics A and B that together inhibit the growth of the microbial strain. Those on the left represent noninhibitory concentrations. The degree of synergism or antagonism is indicated by the shape of the curve with respect to the dashed line, which represents the theoretical situation of simple additivity.

2.2.5.1. Composition of the Medium

Let us take as a simple example an antibiotic that acts by inhibiting the biosynthesis of an amino acid. If tested in a medium without that amino acid, it will appear very active. If tested in a medium containing

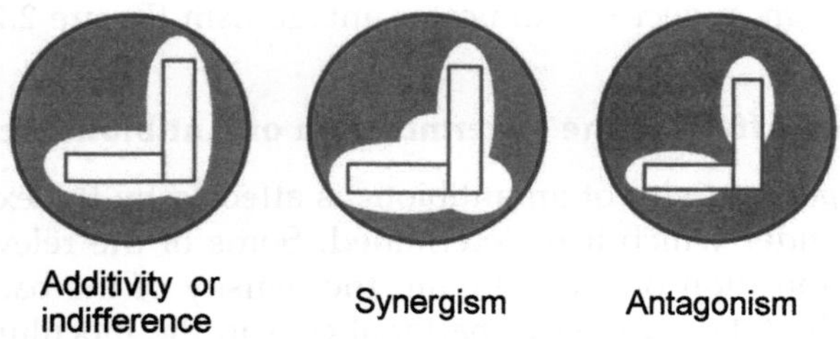

Figure 2.7. Interaction between two antibiotics in solid medium. Each of the two paper strips is soaked with one of the two antibiotics. The shape of the inhibition halo indicates the type of interaction.

Table 2.2. Factors Affecting the Determination of Activity of Some Representative Antibiotics

Antibiotic	Factors
Aminoglycosides	pH; bivalent cations (Mg^{2+}, Ca^{2+})
Penicillins	Dimension of the inoculum
Tetracyclines	pH; bivalent cations (Ca^{2+})
Polyenic macrolides	Sterols
Trimethoprim	Thymine, glycine, methionine
Polyethers	Monovalent cations (K^+)
Methicillin	Dimension of the inoculum (uneven distribution of the susceptibility)
Sulfonamides	*p*-Aminobenzoate; dimension of the inoculum (carrying of antagonist substances); products from one-carbon-atom metabolism

the amino acid and if the bacterium can take it up, the antibiotic will appear to be inactive.

In addition to such specific effects, some of which are reported in Table 2.2, one often encounters less specific effects of the culture medium not directly correlated with the mechanism of action or with the chemical structure of the antibiotic. For example, an antibiotic will generally appear less active in a rich culture medium, which favors a rapid growth, than in a poor one. Similarly, a high temperature favors, within certain limits, the growth rate. The opposite phenomenon can be observed for those antibiotics that require bacterial growth to exert their action.

A very important factor affecting antibiotic activity is the pH of the culture medium. In addition to minor effects, such as that on the microorganism's growth rate, the pH also has a direct effect on the ability of the substance to penetrate bacterial cells: normally, nonionized substances diffuse better through the cell wall and the cytoplasmic membrane than do ionized substances. Therefore, the pH of the medium, by affecting the degree of ionization of a basic or acidic antibiotic, can directly influence its rate of penetration into bacteria and hence its effectiveness.

The presence of blood serum in the culture medium may affect the antibiotic activity. In fact, many antibiotics bind to serum proteins (essentially to albumins) and this decreases the number of free molecules that are available to enter into the bacterial cell, thus generating higher MIC values.

It is noteworthy that with very active antibiotics (MIC values of nanograms), particularly when lipophilic, a high percentage of molecules are absorbed by the test tube wall, especially if the tubes are plastic. Thus, the measured activity appears lower than the real one. This fact may create a paradoxical phenomenon, i.e., the activity appears to increase on addition of even small quantities of serum. In fact, albumin, which is itself absorbed on the tube wall, prevents the absorption of the antibiotic. This phenomenon can be avoided by the use of siliconized tubes.

2.2.5.2. Density and Size of the Bacterial Inoculum

The *density of the bacterium inoculum* is the number of cells inoculated divided by the final test volume, usually expressed as number of viable cells per milliliter of culture. The *size of the inoculum* is the total number of bacterial cells inoculated.

In general, the MIC is not significantly affected by variation in inoculum density in the range commonly used of 10^3–10^6 cells/ml. In fact, even when very low concentrations of antibiotic are used, the ratio of the number of molecules to the number of bacteria is very high (0.01 μg/ml of an antibiotic with a molecular mass of 1000 corresponds to about 10^{12} molecules/ml). However, there are some exceptions. For example, when a large number of antibiotic molecules are absorbed on the outer surface of the bacterial cell and the bacterial density is high, the free molecules available to penetrate into the cells are reduced to a number insufficient for inhibition. Often a large number of antibiotic molecules are necessary to inhibit the growth of a single cell. Most importantly, some bacterial species produce and excrete into the culture medium enzymes that can destroy the antibiotic (e.g., β-lactamases, which can inactivate β-lactam antibiotics: see Chapter 4). The amount of antibiotic destroyed is essentially a function of the enzyme concentration in the culture medium, which, in turn, depends on inoculum density.

At first glance it would appear not to make any difference, once the density of the bacteria has been fixed, whether the test is carried out in a small volume, as, for example, 100 μl in the miniaturized systems, or in the traditional volume of 1 or 2 ml in laboratory test tubes. It would not make any difference if all the members of the bacterial population were perfectly homogeneous and if there were not always a certain degree of variability from cell to cell in their susceptibilities toward the antibiotic. The extent of this variability is clearly a function of the total number of bacteria present in the inoculum. All the susceptible cells will be inhibited, but the less susceptible ones (in theory, even a single cell) will multiply and, after the usual incubation, will generate a dense population of bacteria.

When the frequency of mutants resistant to a specific antibiotic (see Chapter 4) is high, and an inoculum of 10^6 bacteria or more is used, the phenomenon of the *skips* can be observed, i.e., the occasional growth in tubes containing an antibiotic concentration higher than the MIC. This is the result of the random emergence of resistant mutants in that tube.

2.2.5.3. Specific Factors Affecting the Activity in Solid Medium

In addition to the general factors already described (composition of the medium and inoculum), the determination of the activity in solid medium is influenced by specific factors. Only the most important ones are examined here.

The MIC values determined in solid medium may markedly differ from those determined in liquid medium. Agar with its SO_3 groups and its polysaccharide chains can alter the diffusion rate of the antibiotic, of dissolved oxygen, and of nutrients. In addition, agar normally contains a certain amount of bivalent cations. It is not surprising, therefore, that the MIC of an antibiotic against a given bacterium will differ according to whether it is determined in liquid or solid medium, even when the two media are the same except for the agar. It must also be kept in mind that the physiology of the cells may differ when they grow individually in liquid and when they grow in colonies on a solid surface.

It is worth noting that the inhibition halo is determined by the diffusion rate of the antibiotic in the agar medium. The limit of the growth area, and therefore the diameter of the inhibition halo, is in fact determined by the antibiotic concentration at the borderline. The diffusion rate of an antibiotic depends on its physicochemical characteristics and, generally, is inversely proportional to its molecular mass. It depends also on the shape of the molecule, its lipophilicity, and its ionization properties.

For the determination of the susceptibility with the diffusion method it is obviously very important that all the parameters be rigorously standardized, i.e., the culture medium, the quality of the nutrients, and the thickness and concentration of the agar. However, the critical factor is the plate inoculum, as this can markedly influence the diameter of the inhibition halos. For this reason, the inoculum density must be rigorously standardized. Some methods prescribe the use of an inoculum able to give a growth of confluent colonies, while others prescribe inocula giving a growth of semiconfluent colonies, as, in this case, imperfections in the preparation are more evident because of the presence of colonies that are too spread out or, on the contrary, confluent.

2.3. Determination of the Concentration of Antibiotics (Assay)

2.3.1. Chemical and Nonmicrobiological Assays

The *quantitative determination* of an antibiotic can be performed by chemical and physicochemical methods, which are based on characteristics linked to their chemical structures. These methods have experienced an enormous improvement in the last few years. In particular, the ones that have become most common are those that take advantage of the absorption of visible and ultraviolet light, and those based on chromatographic procedures coupled to quantitative detectors (HPLC). For the assays of antibiotics at low concentrations in biological fluids, immunological and enzymatic methods have been recently introduced, e.g., the *radioimmunological assay* (RIA) and the *enzyme-linked immunosorbent assay* (ELISA). Considering the scope of this book, only methods based on the microbiological activity are here described.

2.3.2. Microbiological Assays

2.3.2.1. Agar Diffusion Assay

This method is commonly used for the determination of the concentration of an antibiotic in solution. Traditionally, it consists of placing filter paper disks soaked with the antibiotic solution to be assayed, on the surface of an agar medium containing a dilute suspension of bacteria. After an appropriate period of incubation, the surface of the agar acquires a turbid appearance as a result of the light diffraction caused by the bacterial growth. However, transparent halos will remain around the filter paper disks because of the antibiotic diffusion into the agar. When carried out under standardized and rigorous test conditions, the diameter of the inhibition halo is a function of the logarithm of the antibiotic concentration. A calibration curve can be prepared with solutions at known antibiotic concentration, and the concentration of the unknown solutions can be read from it. In rigorous terms, the logarithm of the antibiotic concentration is proportional to the diameter of the inhibition halo subtracted by the disk diameter. However, within limited concentration intervals, one can obtain a calibration curve even by using the total diameter (see Figure 2.8).

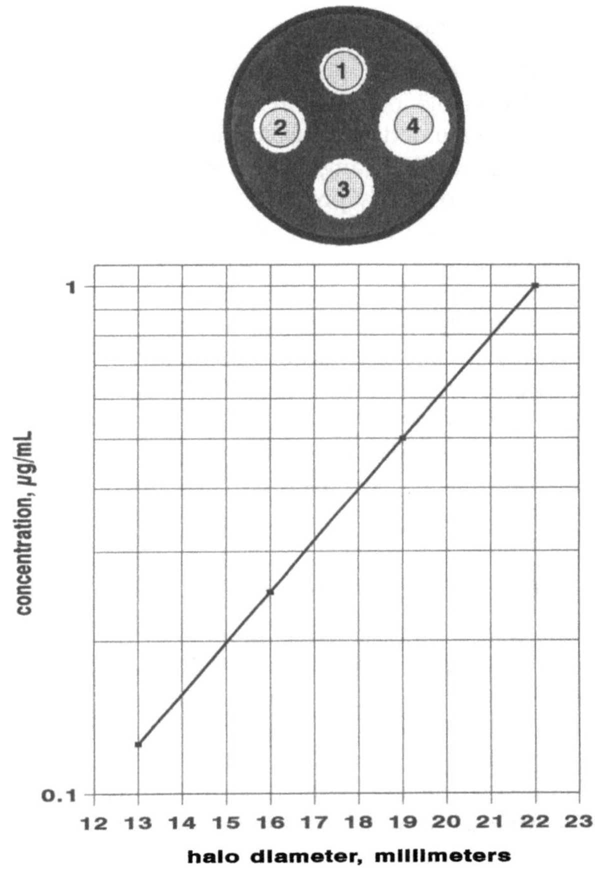

Figure 2.8. Construction of the standard inhibition curve of *Staphylococcus aureus* by rifampin. Discs 1, 2, 3, and 4 were soaked with equal volumes of rifampin solutions with concentrations of 0.125, 0.25, 0.5, and 1 µg/ml and were then placed on the surface of agar seeded with *S. aureus*. The relationship between the logarithms of the concentrations to the diameters of the inhibition halos is shown by the straight line.

An important factor influencing the halo's diameter, beside the concentration, is the absolute quantity of antibiotic present. To obtain a better accuracy in this assay, it is convenient, instead of using disks, to add measured quantities of the antibiotic solutions to small wells, dug in the agar by an appropriate punch. In this way, the larger quantity of solution provides uniformity in the test. In the past, ceramic cylinders of the type used for electric current insulators were used as containers for the antibiotic solutions.

2.3.2.2. Turbidimetric Assay

This is based on the principle that concentrations of an antibiotic lower than the MIC cause a reduction in the growth rate of a bacterial culture, and can be evaluated by a turbidimetric measurement. Within a predetermined concentration range, the logarithm of the optical absorbance of the culture is inversely proportional to the concentration of the antibiotic. This allows us, by the use of solutions of known concentration, to construct a calibration curve and to determine, by comparison, the concentration of the solution of the antibiotic under examination.

This method can also be used for an approximate evaluation of the susceptibility of a given bacterial strain to an antibiotic. In this case, one uses a known concentration of the antibiotic and the susceptibility of the strain is evaluated on the basis of the optical absorbance of the culture after a predetermined period of time. Obviously, a lower value of optical absorbance corresponds to a lower growth and thus to a greater susceptibility of the strain.

2.3.3. Factors Affecting the Determination of the Concentration

All the factors affecting the determination of the antibiotic activity, discussed in the preceding sections, are relevant also for the determination of its concentration, and some of them have particular importance. It has been pointed out, previously, that in the determination of the antibiotic concentration by the agar diffusion method the critical factor influencing the size of the halo is the inoculum density. For this reason, it is essential to alternate on the same plate the disks or the solutions of the products under examination with those of the standards. Another important factor, already mentioned, is the diffusion rate of the antibiotic. In the case of slowly diffusing antibiotics it is convenient to keep the inoculated plate at low temperature for a certain time (e.g., 1 h), to allow an adequate diffusion before activating the growth of the microorganism at its optimal temperature.

The other parameters to be taken into consideration in the quantitative determinations are:

1. The volume of the antibiotic solution. If the volume is insufficient, the concentration in the well (or in the paper disk) can significantly vary with time because of diffusion. As the diffusion rate is in turn a function of the concentration gradient between the well and the agar, this can cause variations in the halo's diameter.
2. The thickness and the concentration of the agar, as the halos are

larger when the former are lower. It is then necessary, in the preparation of large plates, to ensure that these be perfectly leveled in order to have a uniform thickness.

In reference to the turbidimetric method, it is worth noting that the growth rate is very dependent on the temperature, i.e., small differences of temperature among the tubes can lead to large errors, both in the determination of the concentration of antibiotics and in that of the susceptibility of the strains.

The factors described above are only some of the many that can affect the activity of an antibiotic against a given bacterial strain. Thus, it is essential to specify with precision and to standardize all assay conditions when inhibition of bacterial growth is used for quantitative determination of the activity of an antibiotic so that reproducible data from laboratory to laboratory can be obtained.

2.4. Miniaturization and Automation

The use of microbiological analytical techniques has grown tremendously in recent years. Controls of sterility have become ever more frequent, not only in the pharmaceutical and food industries, but also in chemical and electronic component industries. In addition, the preoccupying phenomenon of the spread of bacteria resistant to several antibiotics has caused an enormous increase in the number of requests for antibiograms made to the microbiological laboratories of hospitals. As a consequence, attempts have been made to develop microbiological techniques that would be less labor intensive and less costly. These attempts have developed in two directions: miniaturization and automation.

2.4.1. Miniaturization

Traditionally, tests have been carried out in 1–2 ml of medium, but at present tests are usually carried out in much lower volumes, e.g., 100 μl. Very common is the use of plastic plates, called *microtiters*, each containing 96 wells of 200 μl. The medium, the inoculum, and the antibiotic solutions can be distributed in the wells by appropriate multiple microdilutors, thus saving a great deal of material and labor.

2.4.2. Automation

Automatic systems have been developed for reading inhibition halos or for counting colonies. They consist of optical readers coupled to computerized systems that record and analyze the data. The major development of automatic systems has been devoted to the determination of antibiograms in hospital laboratories. Nowadays, the most frequently used apparatuses are based on systems for microdiluting samples in liquid medium (the volume varies from 0.02 to 0.1 ml). The antibiotics are generally contained, in a dry state, in appropriate tubes or plates. In some cases the antibiotic is absorbed on a paper disk from which it is eluted by contact with the culture medium. Bacterial growth is generally evaluated by automatically measuring the optical density at given time intervals. The elaboration and the interpretation of data are also performed automatically. The results are usually expressed in the form of categories of susceptibility (S = sensitive, I = intermediate, R = resistant), or, sometimes, as MIC.

Chapter 3

The Mechanism of Action of Antibiotics

The information and the concepts discussed in this chapter have been organized using three approaches: (1) systematic, (2) experimental, and (3) with examples.

Systematic approach: antibiotics are divided into large groups based only on what metabolic functions they exert their inhibitory action upon (transcription, replication, translation, cell wall synthesis, membrane synthesis and function, and antimetabolites), and not their chemical structures.

Experimental approach: the study of an antibiotic's mechanism of action involves identifying the bacterial structure with which it interacts, the nature of interaction, and the antibiotic's effect on cell metabolism. Instead of describing these three aspects for each antibiotic, we have chosen to present, in a simplified form, the types of experiment and the interpretation of the results that contribute to the understanding of these elements.

Exemplifying approach: rather than presenting a complete list of all the antibiotics' mechanisms of action, we have preferred to give a few examples for each of the main mechanisms. Thus, representative experimental results are reported for the most relevant antibiotics of each group. Less important antibiotics are also discussed when they illustrate a particular mechanism of action.

Since knowledge of the mechanism of action at the molecular level facilitates understanding the selectivity of action of antibiotics, this concept is also discussed in this chapter.

3.1. General Aspects

An antibiotic is an inhibitor of microbial population growth. Population growth results from reproduction of individual cells, i.e., from duplication of cellular material and subsequent division of the cell into two daughter cells. For an antibiotic to affect the metabolism of a microbial cell, it must

1. Enter the cell and reach the site of action
2. Bind physically to a cellular structure (*target molecule*) involved in a process essential for maintenance of cell growth or homeostasis, and
3. Markedly inhibit this process

At the cellular level, as already described in Chapter 2, an antibiotic can be bactericidal or bacteriostatic. The bactericidal effect may result from either of two mechanisms:

1. The antibiotic interaction with its target molecule induces an irreversible disruption of the cell integrity or functions.
2. It binds to an essential enzyme or cell structure with such a high affinity that the binding is practically irreversible, as when it forms a covalent bond. The bacteriostatic effect occurs when the antibiotic binds with a lower affinity to its target molecule so that when the antibiotic is removed from the bacterial environment the complex dissociates and the target molecule can resume its function.

At the molecular level, antibiotics can inhibit cell growth by interfering with any essential metabolic process. Thus, there are antibiotics that block the energy-providing systems, others that interfere with the synthesis of intermediate metabolites or with the polymerization processes, either by inhibiting them or by altering the genetic information. The specific mechanism by which inhibition takes place is the *mechanism of action*, or mode of action, of the antibiotic. Antibiotics can be classified according to the process with which they interfere. They are commonly divided into the following groups:

1. Inhibitors of cell wall synthesis
2. Inhibitors of the replication or transcription of genetic material
3. Inhibitors of protein synthesis
4. Inhibitors of cell membrane functions
5. Antimetabolites

3.2. Methods of Study

To uncover an antibiotic's mechanism of action, a series of experiments can be carried out whose results will give information at three different levels: (1) in the intact cell, (2) in a partially purified cell-free system, and (3) in one or more purified enzyme systems. Generally, passing from the cellular level to the subcellular ones produces more precise information on the molecular target.

The types of experiments indicated above require the use of the techniques of classical biochemistry and molecular biology. In addition, detailed and precise information on mechanisms of action can be obtained by the use of genetic and recombinant DNA methods.

When the structure of the product under study is similar to that of a known antibiotic, it can be assumed that the new antibiotic has the same or a similar mechanism of action. If the new antibiotic shows a structural similarity to an intermediate of the bacterial metabolism, a reasonable hypothesis is that it acts as an antagonist of that intermediate.

3.2.1. Study of the Activity in Intact Cells

A very useful procedure is that of adding the antibiotic to growing cultures of a susceptible strain and observing the effects on macromolecular synthesis, i.e., DNA, RNA, proteins, and peptidoglycan. These syntheses can be easily followed by adding radioactively labeled specific precursors to the culture medium.

Ordinarily, labeled thymine is used to follow the synthesis of DNA, labeled uracil for RNA, labeled phenylalanine (or another amino acid) for proteins, and labeled acetylglucosamine for peptidoglycan. At regular intervals one measures the radioactivity incorporated into the corresponding macromolecule by the bacteria, which is an index of the rate of synthesis. On addition of an antibiotic, inhibition of the synthesis of one class of macromolecules may be observed, indicating which metabolic pathway is affected by the antibiotic. Naturally, stopping the synthesis of a macromolecule essential for cell growth affects all other cellular functions, with the eventual arrest of the synthesis of other types of macromolecules. Therefore, to establish what the primary effect is, one must observe the time course of the various events. The primary effect is usually that seen earliest, but there can be simultaneous arrest of synthesis of more than one or even all macromolecular systems. For example, this is the case when the antibiotic acts on the respiratory function (with consequent block of ATP synthesis, necessary for the synthesis of all types of macromolecules) or on the integrity of the cell membrane.

Figure 3.10 in Section 3.4.2 shows the kinetics of macromolecular synthesis in a growing bacterial culture in the presence or absence of rifampin. Following addition of the antibiotic, a rapid block in the incorporation of uracil is observed, followed within a few minutes by inhibition of phenylalanine incorporation. One can deduce that the primary effect is inhibition of RNA synthesis, with the consequent arrest of protein synthesis. DNA synthesis appears to be affected only slowly, suggesting that RNA synthesis inhibition is not a consequence of the antibiotic binding to DNA.

3.2.2. Study of the Activity in Partially Purified Cell-Free Systems

Once the primary effect of an antibiotic has been determined, i.e., which macromolecular synthesis is inhibited first, one must ascertain whether the inhibition is the result of interference with (1) the synthesis of precursors or their activation, (2) the enzymes or organelles involved in polymerization, or (3) the information system that determines the order in which the precursors are incorporated into the polymer.

This type of study is usually carried out with enzyme systems obtained from partially purified cell homogenates that are able to polymerize the monomers into macromolecules. Cell-free systems for *in vitro* synthesis of proteins, nucleic acids, and peptidoglycan can be prepared. If the antibiotic inhibits *in vitro* the synthesis of the same macromolecule inhibited in the growing cell, then its action is on the polymerization process, or possibly on the information molecules. If it does not, one should look for its effect on synthesis of precursors or on their activation.

To return to the example of rifampin, which inhibits RNA synthesis in the intact cell, one must determine its ability to inhibit, *in vitro*, the activity of a transcription system that contains among other things DNA, nucleotide triphosphates, and polymerizing enzymes. As rifampin inhibits the activity of this system when catalyzed by bacterial RNA polymerase but not by RNA polymerase from eukaryotes, one can deduce that the block of RNA synthesis observed in growing cells is caused by a direct interference with the enzyme RNA polymerase.

3.2.3. Study of the Activity in Purified Enzyme Systems

The polymerizing systems mentioned above usually contain many components, any one of which might be the target of the antibiotic's effect. Within certain limits one can identify which enzymatic reaction is affected and determine the component on which the antibiotic acts. Sometimes, this can be obtained by identifying the formation of a complex between the antibiotic and the enzyme protein, either by radiochemical methods or by detecting variations in the chromatographic or electrophoretic mobility of the protein in the presence of the antibiotic. In other cases, it is possible to isolate, from mutants resistant to the antibiotic, a strain whose resistance results from an alteration of the protein on which the antibiotic acts. To identify it, one can operate *in vitro* with reconstituted systems where one component at a time is replaced by that from the resistant mutant. Lack of inhibition of one of the combinations identifies the component involved in antibiotic activity.

Let us discuss again the example of rifampin: a number of mutants resistant to this antibiotic have been isolated. The RNA polymerase of some of these is extracted and used in the *in vitro* systems described above. Since under these conditions no inhibition of synthesis is observed at low antibiotic concentrations, one can deduce that the mutation to resistance is the result of an alteration of RNA polymerase, thereby confirming that this enzyme is indeed the target of the antibiotic action. RNA polymerase comprises several different subunits. These subunits can be separated and the enzyme can be reconstituted by utilizing a different subunit deriving from the resistant mutant each time. By determining antibiotic activity on the various reconstituted polymerases, one can identify the subunit that confers resistance. For rifampin, it was shown that the subunit β was responsible. A similar conclusion has been reached by the observation that rifampin is able to form a complex with this subunit.

3.2.4. Genetic and Recombinant DNA Methods

Genetic techniques can be used to obtain preliminary information on a new antibiotic's mechanism of action. For example, if cross-resistance has been ascertained between the new antibiotic and a known antibiotic, it is probable that they have similar mechanisms of action. If there is no cross-resistance, another site of action is implied. This information then suggests appropriate biochemical experiments to confirm the hypothesis.

The mechanism of action can be determined by genetic methods when resistant mutants are available whose resistance is caused by a mutation of the target enzyme. The procedure consists of preparing in a suitable vector a genetic bank of the DNA of the resistant mutant and of transferring, by genetic engineering techniques, the DNA bank into a sensitive strain. By isolating the clones in which resistance has been transferred, one can establish the responsible gene and thus identify the corresponding protein. Once the responsible gene has been identified, the use of genetic engineering methods allows us to establish precisely the molecular site with which the antibiotic complexes. In fact, one can compare the sequence of the gene of the resistant mutant with that of the sensitive strain, and then deduce which alterations in the amino acid sequence are responsible for resistance.

Through this type of approach it was possible to show that resistance to rifampin usually arises from mutation of certain amino acid residues located between position 507 and position 533 of the subunit β of RNA polymerase.

3.3. Inhibitors of Cell Wall Synthesis*

Before discussing the mechanisms of action of inhibitors of cell wall synthesis, we shall briefly describe the chemical structures of the cell wall of different microorganisms and the pathways through which the main components are synthesized.

3.3.1. Structure and Architecture of the Cell Wall

The cell wall is a rigid structure that envelopes the microbial cell, determines its shape, and protects it from bursting as a consequence of the high internal osmotic pressure. In prokaryotes, the basic structure consists of a complex three-dimensional network composed of *peptidoglycan* (a glycopeptide also called murein) and other polymers (e.g., polysaccharides, lipoproteins) that vary in different bacterial species. Peptidoglycan consists of long polysaccharide filaments formed of two alternately arranged monomers, *acetylmuramic acid* (M) and *acetylglucosamine* (G), connected by β,1-4 glycosidic linkages. The filaments are covalently interconnected by peptide chains that branch off from the acetylmuramic unit giving rise to the three-dimensional structure (Figure 3.1). The interconnected peptide chains are different in each bacterial species.

*The chemical structures of the antibiotics in this and the following sections, when not shown, can be found in Chapter 5.

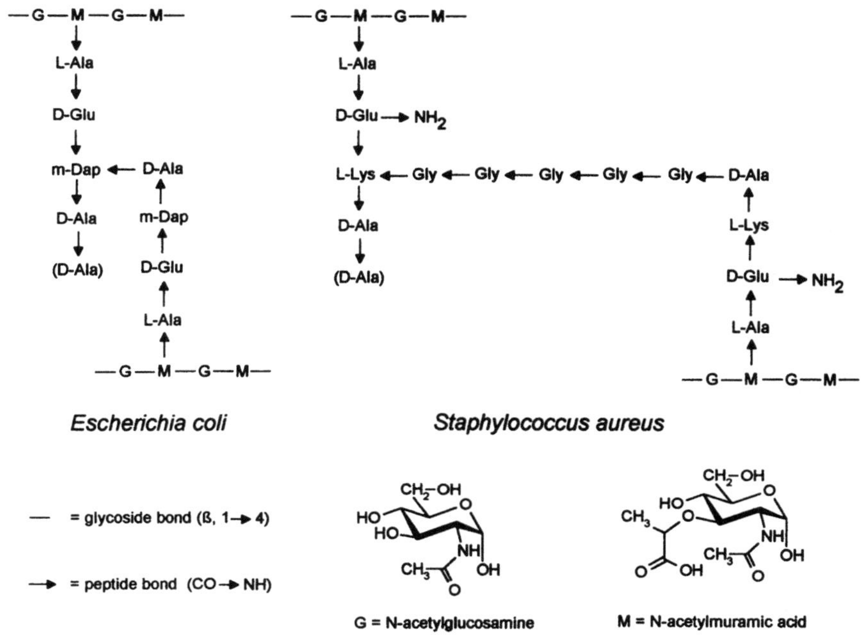

Figure 3.1. Fragments of the primary structure of peptidoglycans of *Escherichia coli* and of *Staphylococcus aureus*.

In some cases, tetrapeptides stemming from two polysaccharide chains are connected directly (e.g., *E. coli*), in others the linkage is mediated by an additional amino acid or a short peptide chain (e.g., *S. aureus*). In addition to the three-dimensional network of peptidoglycan, there are other structures in the bacterial cell wall that differ from organism to organism, but which are basically of two types: those of gram-positive and those of gram-negative bacteria.

3.3.1.1. Gram-Positive Cell Wall

This consists of a layer that appears uniform when viewed under the electron microscope and is made up of peptidoglycan, proteins, and considerable quantities of teichoic acids. In all gram-positive bacteria there is a "membrane teichoic acid," so called because it is situated between the wall and the cytoplasmic membrane. It consists of a chain of alternating glycerols and phosphates. In addition, there are "wall teichoic acids" which are bound covalently to the peptidoglycan. They may have a structure similar to that of membrane teichoic acids or may contain ribitol instead of glycerol. In some rare forms of teichoic acid, the basic polymer contains a sugar such as glucose or *N*-acetylglucosamine in addition to glycerol.

3.3.1.2. Gram-Negative Cell Wall

The cell wall of gram-negative bacteria is thinner but more complex. As shown by the electron microscope, it is composed of a number of distinct layers. The innermost layer is a peptidoglycan where the third amino acid residue of the tetrapeptide is normally diaminopimelic acid. The intermediate layer consists of lipoproteins. The peptidoglycan is not covalently bound to the cytoplasmic membrane but to the lipoproteins. The external layer, called the outer membrane, can be described as a double layer similar to the cytoplasmic membrane, but where the inner layer only is composed of phospholipids, while the outer layer is composed of lipopolysaccharides. Across the outer membrane there are hydrophilic pores formed by special proteins called *porins*. The zone between the cytoplasmic membrane and the outer membrane is called the *periplasmic space*. It contains various enzymes, some of which have an antibiotic-inactivating activity.

3.3.1.3. Biosynthesis of Peptidoglycan

The synthesis of peptidoglycan in *E. coli* can be divided into three stages (Figure 3.2).

1. Formation of basic units. UDP-*N*-acetylmuramic acid is synthesized by condensation of *N*-acetylglucosamine (activated as uridine diphosphate at the anomeric carbon) with phosphoenolpyruvate, followed by reduction of the double bond. To UDP-*N*-acetylmuramic acid are successively added L-alanine, d-glutamic acid, and *meso*-diaminopimelic acid. To the resulting intermediate, called muramyl-tripeptide, the dipeptide D-alanyl-D-alanine is added. The latter originates from the isomerization and condensation of two L-alanine molecules. The final product of this series of reactions, which take place in the cytoplasm, is UDP-muramyl-pentapeptide (Figure 3.3).
2. Completion and transfer of monomers through the membrane. The muramyl-pentapeptide is charged on a lipid carrier, undecaprenylphosphate, with loss of UMP and formation of a diphosphodiester between the carrier and the anomeric carbon of muramic acid. The monomer synthesis is completed by the addition of *N*-acetylglucosamine to muramic acid through a β,1-4 gly-

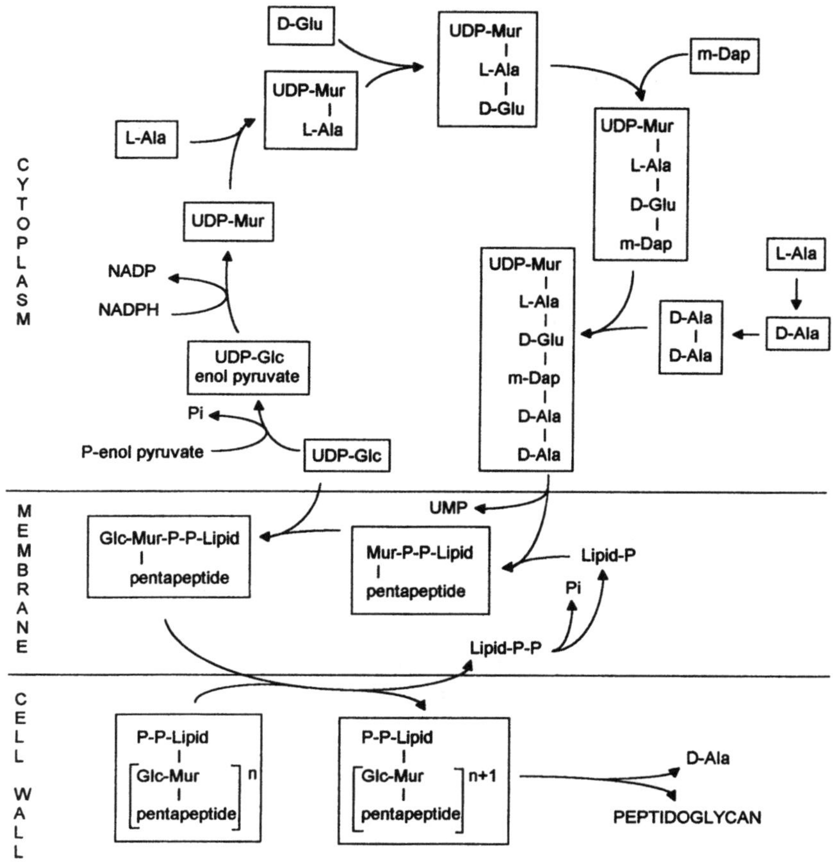

Figure 3.2. Biosynthesis of peptidoglycan. Glc = *N*-acetylglucosamine; Mur = *N*-acetylmuramic acid; L-Ala = L-alanine; d-Glu = d-glutamic acid; m-Dap = *meso*-diaminopimelic acid; D-Ala = D-alanine; Lipid-P = undecaprenylphosphate.

cosidic bond. These reactions take place in the cytoplasmic membrane (Figure 3.2).

3. Assembly of peptidoglycan chains and construction of the three-dimensional structure. The final steps of peptidoglycan synthesis are catalyzed by various enzymes (see penicillin binding proteins in Section 3.3.3.3), designated according to their function:

    a. *Transglycosylases*, which catalyze the formation of a β-glycosidic bond between C-1 of acetylmuramic acid and C-4 of acetylglucosamine

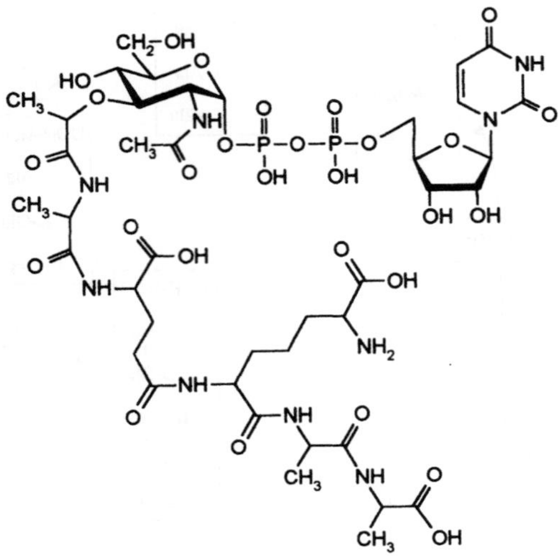

Figure 3.3. Structure of UDP-muramyl-pentapeptide.

b. *Transpeptidases,* catalyzing the formation of peptide bonds between D-alanine in position 4 and the nitrogen at the ϵ position of the diaminopimelic acid of an adjacent pentapeptide (this reaction takes place by releasing the terminal D-alanine of the pentapeptide donor)
c. *D-Carboxypeptidases,* which hydrolyze the terminal D-alanine of the pentapeptides
d. *Endopeptidases,* hydrolyzing peptide bonds in the preexisting peptidoglycan chains

The disaccharide pentapeptide is transferred to an acceptor, a chain of nascent peptidoglycan, by means of transpeptidation and transglycosylation reactions. In this latter reaction undecaprenylpyrophosphate is released and enters the cycle again. During the process of microbial growth and division when new peptidoglycan is made, endopeptidases become active on the internal surface of the wall. These enzymes partially hydrolyze the existing chains, and make free ends available to receive the nascent peptidoglycan chain, again by transpeptidation and transglycosylation reactions (Figure 3.2).

Polysaccharide chains may then be further connected by formation of peptide bonds catalyzed by transpeptidases. This connection is called cross-linking. These reactions take place on the external surface of the cytoplasmic membrane.

In all bacterial species, peptidoglycan synthesis proceeds according to the pattern described for *E. coli*, but some important variations may be found. For instance, in *S. aureus*:

1. The third amino acid in the pentapeptide is lysine rather than diaminopimelic acid
2. When the synthesis of disaccharide-pentapeptide is completed, a chain of five glycine molecules is attached (through peptide bonds) to the ϵ amino group of lysine.
3. The transpeptidation reaction occurs between the carboxyl group of the penultimate D-alanine of a pentapeptide (with release of the terminal D-alanine) and the amine of the terminal glycine.

This scheme of peptidoglycan synthesis applies to the cell wall at various phases of cell growth. At least two of these can be distinguished: one associated with cell elongation and one involved in septum formation. However, others may exist, as demonstrated by the morphological variations shown by bacteria treated with different inhibitors of peptidoglycan synthesis.

3.3.1.4. Fungal Cell Wall

This is made up of polysaccharides that either are structured in fibrils or are amorphous. In filamentous fungi the fibrils are usually made of *chitin* (poly-*N*-acetylglucosamine, β,1-4) and of β-glucan. In yeasts chitin is a minor component, but it plays an important role in morphogenesis. The major component is instead β,1-3 glucan containing some β,1-6 ramifications. Some mannans are also present, which are covalently bound to polypeptides.

3.3.2. Characteristics of Inhibitors of Bacterial Cell Wall Synthesis

These inhibitors can be divided into two groups:

1. Inhibitors of peptidoglycan synthesis
2. Inhibitors of synthesis or assembly of other cell wall components. The vast majority of known cell wall synthesis inhibitors belong to the former group and they can be classified according to the level at which they act, namely:
    a. Inhibitors of cytoplasmic reactions, i.e., the synthesis of muramyl-pentapeptide
    b. Inhibitors of the reactions that take place at the membrane level

c. Inhibitors of the formation of the three-dimensional structure

Regardless of their specific sites of action, most known inhibitors of bacterial cell wall synthesis have several characteristics in common.

They are generally bactericidal. The bactericidal effect can operate via two different mechanisms: lytic or nonlytic. In the first case, mortality is the result of a loosening of peptidoglycan structure which brings about a disaggregation of the cell as a consequence of the high osmotic pressure of the cytoplasm. In the second case, other mechanisms may intervene, such as the irreversible inhibition of septum formation or of cell elongation.

They are generally inactive against resting cells. All the mechanisms of bactericidal activity are mediated by the production of enzymes that are active when the cells are growing but not when they are resting. As a consequence, the cell wall synthesis inhibitors are inactive against bacteria in the stationary phase.

They are obviously inactive against microorganisms that lack a cell wall. Three types of these organisms are known:

1. Mycoplasmas—bacteria occurring in nature, as parasites of animals and plants. They do not have a cell wall and are unable to synthesize it under any conditions.
2. L-forms—bacteria generally obtained under laboratory conditions but sometimes found in nature (e.g., in infections of the urinary tract) that are able to grow and to multiply if maintained under hypertonic conditions. They tend to regenerate a cell wall unless a cell wall synthesis inhibitor is present.
3. Protoplasts—bacterial forms produced in the laboratory, but which, in contrast to L-forms, are unable to multiply. However, under special conditions they can regenerate their cell wall.

3.3.3. Examples of Inhibitors of Peptidoglycan Synthesis

As already mentioned, these antibiotics can be classified according to the level at which they act.

3.3.3.1. Inhibitors of Muramyl-Pentapeptide Synthesis

Among the inhibitors of the early reactions, two antibiotics have been particularly studied.

Fosfomycin. This antibiotic inhibits the first reaction of *N*-acetyl-muramic acid synthesis, i.e., the condensation of UDP-*N*-acetylglucosamine with phosphoenolpyruvate catalyzed by pyruvyl transferase to give enolpyruvyl-*N*-acetylglucosamine. Fosfomycin, which can be considered a structural analog of phosphoenolpyruvic acid, binds covalently to pyruvyl transferase, resulting in its irreversible inactivation.

Cycloserine. The action of this antibiotic, which is particularly active against mycobacteria, is antagonized by addition of D-alanine to the culture medium. This has suggested that cycloserine interferes with D-alanyl-D-alanine synthesis. This has been demonstrated *in vitro*, by the finding that cycloserine inhibits both the enzyme that converts L-alanine into D-alanine (alanine racemase) and the enzyme catalyzing the formation of the peptide bond between two molecules of D-alanine (D-alanyl-D-alanine synthetase). Again, antibiotic action appears to result from the structural similarity between the antibiotic molecule and one particular conformation of D-alanine.

3.3.3.2. Inhibitors of Reactions Occurring in the Membrane

Today, there are no clinically important antibiotics acting at this level. However, worth mentioning are bacitracin, used in animal husbandry, and ramoplanin, presently under development.

Bacitracin. This is a peptide antibiotic isolated from cultures of *Bacillus subtilis*. It inhibits the conversion of undecaprenylpyrophosphate (released during the elongation of the peptidoglycan chains) into undecaprenylphosphate, which is the acceptor of muramyl-pentapeptide in the transfer reaction. This antibiotic is thought to bind to the reaction substrate rather than to the enzyme catalyst. In fact, bacitracin also inhibits some enzymatic reactions whose substrate is a lipid-pyrophosphate, and not other dephosphorylation reactions.

Ramoplanin. This new bactericidal agent is active against gram-positive bacteria. Its action on cell wall synthesis is demonstrated by a rapid and specific inhibition of the uptake of *N*-acetylglucosamine by growing cells and by the resulting cellular accumulation of UDP-muramyl-pentapeptide. Experiments carried out with membrane preparations have shown that ramoplanin inhibits the enzyme (*N*-acetylglucosaminyl-transferase) that adds *N*-acetylglucosamine to undecaprenyl-muramyl-pentapeptide.

3.3.3.3. Inhibitors of Chain Elongation and Formation of the Three-Dimensional Cell Wall Structure

Vancomycin and Teicoplanin. These are the only members in clinical use of the dalbaheptide family of glycopeptide antibiotics. Other members of this group are: avoparcin, used in animal husbandry, and ristocetin, formerly used in clinics, but now withdrawn. Characteristic of this family is the ability to form complexes with peptides terminating in D-alanyl-D-alanine. Consequently, these antibiotics act by inhibiting peptidoglycan synthesis, as demonstrated by their inactivity on L-forms and by the rapid inhibition of the uptake of *N*-acetylglucosamine in growing cells. However, it has not been clearly ascertained what reaction is specifically inhibited, or to which peptidoglycan precursor they preferentially bind. As UDP-muramyl-pentapeptide accumulates in treated cultures and undecaprenyl-disaccharide in membrane systems *in vitro*, one can deduce that the inhibition takes place at the chain elongation or at the cross-linking level. The antibiotic effect is mainly bactericidal but it does not bring about a massive lysis of the cells.

β-Lactam Antibiotics. This family of antibiotics includes the penicillins, the cephalosporins, the cephamycins, and the newer β-lactams such as monobactams and thienamycins. All have a similar mechanism of action, i.e., they interfere with the formation of the cell wall by blocking the final phase of peptidoglycan synthesis (Figure 3.4). However, β-lactams differ in the degree to which they can inhibit the different enzymes involved in the peptidoglycan synthesis, such as transpeptidases, d-carboxypeptidases, and, indirectly, enzymes performing transglycosylation.

In any event, their molecular mechanism can be generally exemplified by the interaction of penicillin G with transpeptidase (Figure 3.5). This enzyme acts in two stages: first, it catalyzes the hydrolysis of the pentapeptide terminal D-alanine with formation of an ester bond between the carboxyl group of the alanine in position 4 and an active serine site on the enzyme; then, it breaks this ester bond with formation of a peptide bond between the alanine carboxyl and the *m*-Dap amino group of an adjacent pentapeptide. Because of its structural similarity to D-alanyl-D-alanine, the antibiotic behaves as a substrate for the first reaction; opening of the β-lactam ring unmasks a carboxyl group that forms an ester bond with the serine hydroxyl group of the enzyme. An analogous mechanism operates in the inhibition of the d-carboxypeptidases that catalyze the first stage of the reaction, but then hydrolyze the ester bond.

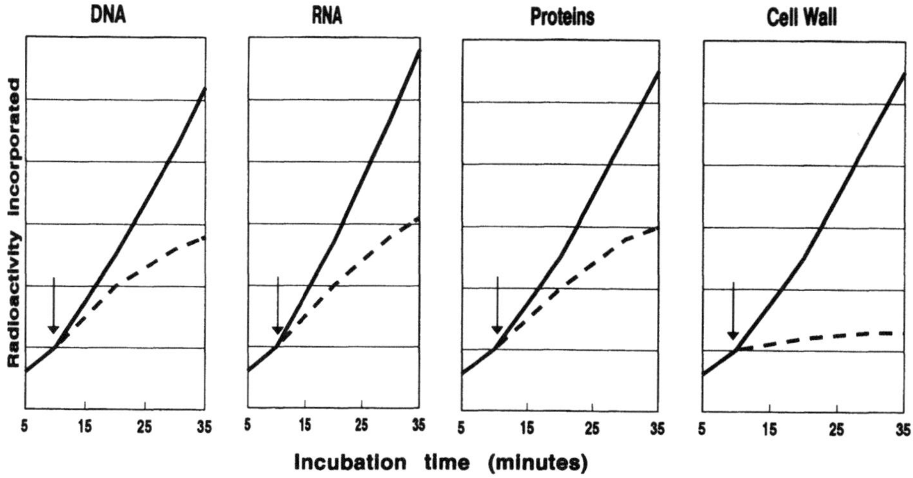

Figure 3.4. Effect of penicillin G on incorporation of radioactive precursors into DNA, RNA, proteins, and cell wall of growing bacteria. The solid line indicates incorporation into a control culture, the dashed line incorporation into a culture to which antibiotic was added at the time indicated by the arrow.

The enzymes inhibited by the β-lactam antibiotics have been identified on the basis of their ability to form a covalent bond with these antibiotics. Accordingly, they have been termed *penicillin binding proteins* (PBP).

PBPs are present in almost all bacteria but they vary from species to species in number, size, and affinity for β-lactam antibiotics. They are localized nonrandomly on the surface of the cytoplasmic membrane. A given organism may contain four to eight PBPs, with molecular weights from 35,000 to 120,000. When a "novel" PBP is identified in an organism, it is numbered as a derivative of the established ones. For instance, the novel 78,000-M_r PBP found in methicillin-resistant *S. aureus* (in addition to the existing PBP-2 of 80,000 M_r) has been designated as PBP-2a (or PBP-2′).

PBPs essential for cell growth normally have transpeptidase activity. These control the fundamental processes of cell elongation and division, and usually have a high molecular weight (60,000 to 120,000 M_r). In a given organism there may be two to four essential PBPs, which are the potential targets of β-lactam antibiotics.

The PBPs of *Escherichia coli* have been extensively studied and their characteristics are shown in Table 3.1. The various β-lactam antibiotics

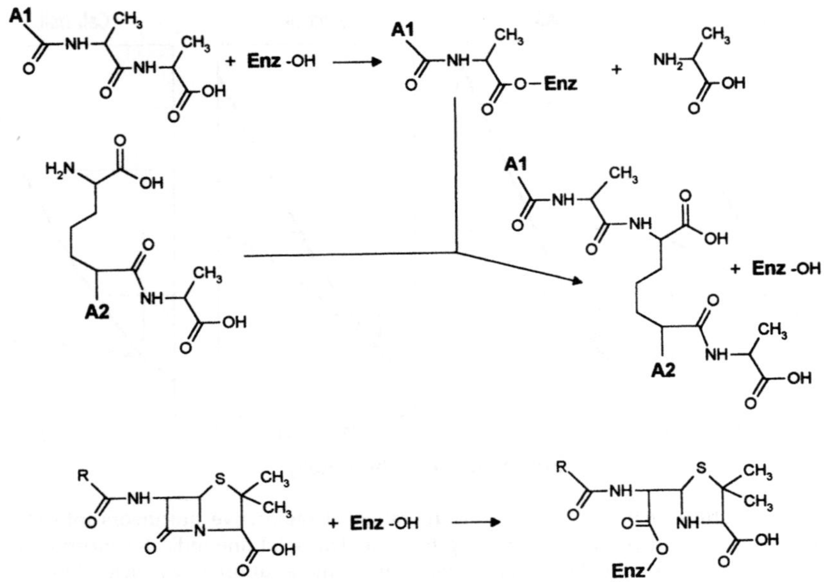

Figure 3.5. Transpeptidation promoted by transpeptidases (represented by Enz-OH) and interference of penicillins with these enzymes. Structural formulas are drawn in a way to emphasize the similarity between penicillins and D-alanyl-D-alanine. A1 and A2 are two peptidoglycan filaments.

Table 3.1. Synopsis of *E. coli* Proteins That Interact with Penicillins (PBP)

PBP	Function	Antibiotics with high affinity
1A and 1B	Transpeptidases and transglycosylases. Extend peptidoglycan chains for cell elongation	Penicillin G and various cephalosporins
2	Transpeptidases and transglycosylases. Initiate insertion of nascent peptidoglycan in the extension sites	Mecillinam and imipenem
3	Transpeptidases and transglycosylases. Specific for septum formation	Cephalexin, other cephalosporins, piperacillins?
4	D-Carboxypeptidases and endopeptidases. Prepare the attack sites for the nascent chains	none known
5 and 6	D-Carboxypeptidases. Determine the degree of cross-linking	Cefoxitin
7	Not known	Imipenem

display different degrees of affinity toward the individual PBPs. The extreme case is that of mecillinam, which binds only to PBP-2. As the various PBPs may be responsible for the maturation of different types of peptidoglycan, the different β-lactams cause a variety of morphological effects. For example, benzylpenicillin has a lytic action, cephalexin inhibits septum formation but not cell elongation and thus gives rise to elongated forms of *E. coli*. Mecillinam inhibits cell elongation but not septum formation, producing rounded cells.

The mechanism underlying the bactericidal effect of β-lactams is very complex. The lethal effect can be manifested with or without cell lysis. The lytic effect is obviously the result of a weakening of the wall structure, which becomes unable to control the swelling caused by the internal osmotic pressure. Formerly, it was hypothesized that this weakening was related to "unbalanced" growth: it was believed that the bacterial autolysins, enzymes that hydrolyze the bonds in peptidoglycan, would act in a continuous way to create sites of attachment for chain growth. In the presence of penicillins, the lack of formation of new interpeptide linkages and the normal action of the autolysins would result in a "weak" structure of the wall. On the contrary, recent investigations have demonstrated that, although some of these autolysins are actually responsible for the lytic effect, they do not operate continuously, but are specifically activated at particular phases of cell growth. The effect of some β-lactams could arise from activation of autolysins at the wrong times or at unsuitable sites, following inhibition of peptidoglycan synthesis. This interpretation explains the lack of a bactericidal effect of penicillins on resting cells. In fact, when the cells are in stationary phase, notably when nutrients are lacking, a mechanism of autoregulation, called the *stringent response*, comes into effect, which blocks the synthesis of various macromolecules including peptidoglycan. It is suggested that the stringent response could also affect the mechanism of autolysin activation involved in peptidoglycan synthesis.

The nonlytic effect can take place either in bacteria lacking some autolysins or in those in which the inhibition of one PBP does not induce activation of an autolysin. The final effect can be bacteriostatic only (in this case the strains are said to be "tolerant" toward the bactericidal action of β-lactams), or bactericidal, when the block of cell wall construction is irreversible.

3.3.4. Inhibitors of Outer Membrane Synthesis

Two antibiotics are known whose mechanism of action appears to involve alteration of the outer membrane of gram-negative bacteria.

The action of the antibiotic *bicyclomycin* on the outer membrane was essentially deduced from the observation that it inhibits only gram-negative bacteria and induces morphological changes at their surface. Inhibition of biosynthesis of wall lipoproteins and of their bonds with peptidoglycan has been hypothesized. However, recent evidence based on genetic experiments strongly indicates that bicyclomycin acts by inhibiting the function of the termination factor ρ in RNA synthesis. The relationship of this inhibition to the alteration of cell surface morphology remains unclear.

The antibiotic *globomycin* also is active uniquely against gram-negative bacteria. Alteration of the outer membrane appears to be the result of inhibition of the maturation process of lipoproteins, which are essential components of the cell wall of gram-negative bacteria.

3.3.5. Inhibitors of Fungal Cell Wall Synthesis

3.3.5.1. Inhibitors of Chitin Synthesis

Polyoxins and Nikkomycins. This group of antibiotics possess the general structure of nucleosides substituted in position 5′ by one or more amino acids. They are particularly active against filamentous fungi, moderately active against yeasts, but inactive against bacteria. They competitively inhibit the enzyme chitin synthetase, most likely because their structure resembles that of the enzyme substrate, UDP-*N*-acetylglucosamine. The low activity against yeasts seems to be related to the minor importance of chitin in the wall of these microorganisms and to poor permeation. In fact, they are transported into the cells by the permeases that carry dipeptides, and consequently they must compete with these products for penetration.

3.3.5.2. Inhibitors of Glucan Synthase

Echinocandins and Papulacandins. These two groups of antibiotics, although of quite different chemical structure, share the same mechanism of action: inhibition of the enzyme β,1-3 glucan synthase of different yeasts and fungi. However, activity on living cells is observed only with yeasts, particularly *Candida albicans*.

3.4. Inhibitors of Replication and Transcription of Nucleic Acid

3.4.1. Replication and Transcription of Genetic Information

The synthesis of nucleic acids can be distinguished in two phases: (1) synthesis of precursors (nucleotides and deoxynucleotides) from in-

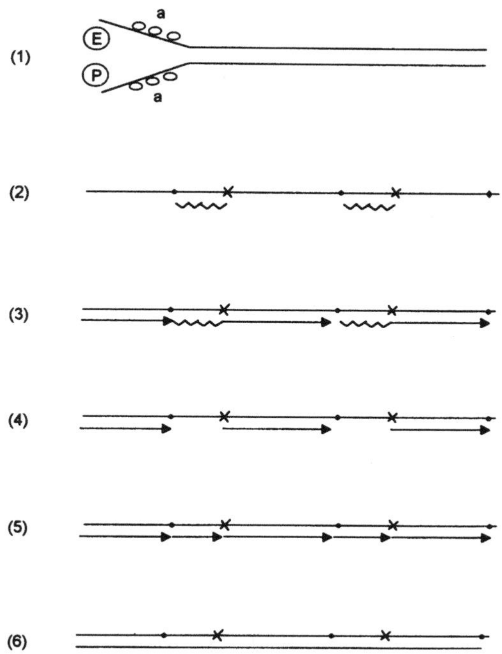

Figure 3.6. Schematic model of DNA replication in *E. coli*. (1) The enzyme E and ATPase P unfold the double helix. The Alberts protein (a) keeps the two strands separated. (2) A specific RNA polymerase synthesizes short RNA chains. (3) DNA polymerase III synthesizes polydeoxynucleotides using 3′-OH of RNA as primer. (4, 5) DNA polymerase I degrades the RNA fragments and completes the DNA synthesis. (6) A ligase joints the adjacent fragments.

termediate molecules of the cell metabolism and (2) enzymatic polymerization of nucleotides to form a macromolecule whose sequence is determined by the sequence of the bases in the DNA template.

Several different enzymes are involved in the replication of DNA. A simplified scheme for DNA replication in prokaryotes is shown in Figure 3.6. The following steps can be identified:

1. The strands of the double helix are unfolded by the action of enzymes called helicases and are kept unfolded by the action of a specific protein (formerly called Alberts protein). A topoisomerase relieves the superhelicity induced by unfolding by transiently breaking one strand.
2. A specific RNA polymerase, called primase, synthesizes short chains of RNA on each strand, at positions corresponding to specific initiation sites.
3. The replicative DNA polymerase (polymerase III in *E. coli*) synthesizes small DNA pieces, called Okazaki fragments, starting from the 3′-OH of RNA, which functions as primer.
4. Another enzyme (in *E. coli* the DNA polymerase I) degrades the RNA fragments, and

5. Simultaneously replicates the sequences between the Okazaki fragments.
6. A ligase joins the adjacent fragments together.

When replication of the circular chromosome has been completed, the progeny DNA molecules are interwound (concatenation). The separation is accomplished by another topoisomerase, which cleaves both strands of a double helix and rejoins them after the other double helix passes through the created interruption. By an analogous mechanism, DNA gyrase introduces supercoils into the daughter DNA chains to make them more compact.

The process of RNA synthesis is less complex. In bacteria it is catalyzed by RNA polymerase, an enzyme composed of four proteins, α, α, β, β', and one factor, σ, essential for the recognition of the DNA sequences specifying transcription initiation. Functions of RNA polymerase are:

1. To separate the DNA strands by formation of a complex with DNA
2. To place the first nucleotide in the correct position on one DNA strand (initiation)
3. To incorporate the second nucleotide and form the first phosphodiester bond between the first and second nucleotides (completion of this reaction signals the end of initiation)
4. To move along the DNA strand and continue to incorporate nucleotides joined by phosphodiester bonds (elongation)
5. To terminate the process when a particular sequence of DNA has been reached (termination). In many cases the termination involves another protein, rho factor (ρ).

In eukaryotes there are three RNA polymerases, each being composed of several proteins, which synthesize specific types of RNA. RNA polymerase I, localized in the nucleolus, synthesizes 28 and 18 S ribosomal RNAs, RNA polymerase II synthesizes messenger RNAs, and RNA polymerase III makes transfer RNAs and 5 S ribosomal RNA.

3.4.2. Examples of Replication and Transcription Inhibitors

Replication and transcription inhibitors can be divided into two groups: those that inhibit the synthesis of precursors and those that inhibit nucleic acid polymerization. The latter can in turn be classified into inhibitors of the template function of DNA and inhibitors of enzymes, such as gyrase, DNA polymerase, or RNA polymerase.

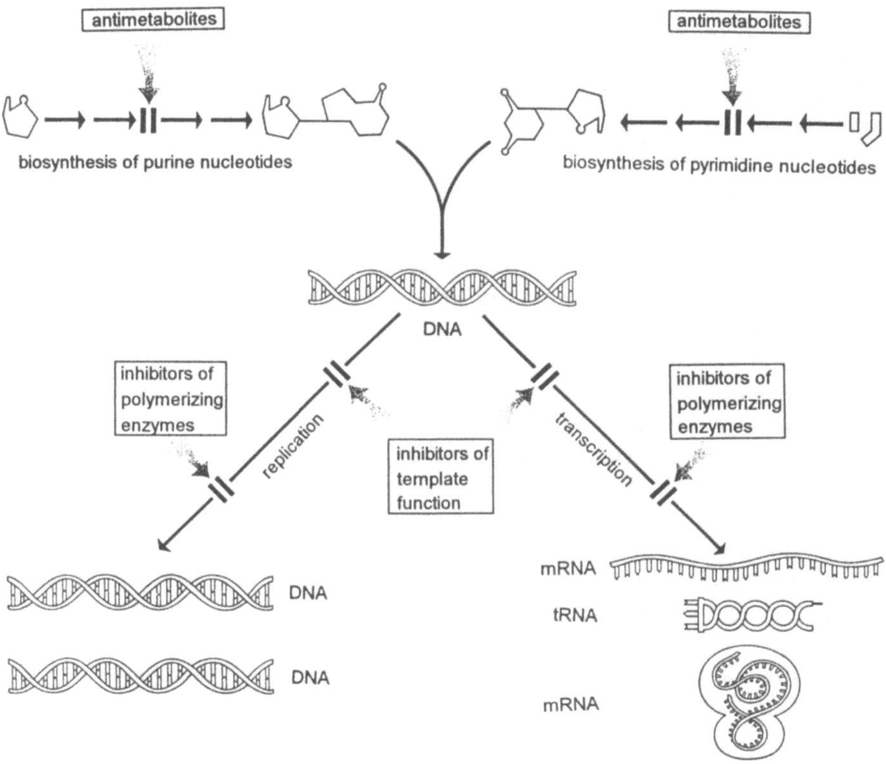

Figure 3.7. Summary diagram of mechanisms of action of nucleic acid synthesis inhibitors.

Precursor analogues, which may interfere with either nucleotide synthesis or polymerization, are discussed in Section 3.7 as "antimetabolites." Inhibitors of DNA template function and of enzymes are discussed below (Figure 3.7).

3.4.2.1. Inhibitors of the Template Functions of DNA

Many substances inhibit replication and transcription by interfering with the template function of DNA, either by forming a nonfunctional complex (because of disruption of the tertiary structure) or by causing chemical alterations in the structure (nicked strands, scission of bases, formation of covalent bonds between the two strands).

These antibiotics have two characteristics in common: (1) They bind indiscriminately to the DNA in a variety of prokaryotic and eukaryotic cells. Thus, they are not specific in their effect and are generally very toxic. (2) They inhibit the synthesis of DNA and RNA simultaneously, although under certain conditions one type of nucleic acid is preferentially inhibited.

1. *Inhibitors that chemically modify DNA.*

- *Mitomycins.* These cytotoxic agents bind covalently to DNA, forming bridges between the two strands and thus preventing their separation. It is noteworthy that this reaction does not occur *in vitro* because to be active mitomycins must be reduced as a result of cellular metabolism. Because they covalently bind DNA the effects are irreversible and their action is bactericidal. As a result of their lack of selectivity they are very toxic, and cannot be used as antimicrobial agents. However, they are of interest as antitumor agents.
- *Bleomycins.* These, like the analogous *phleomycins,* are DNA-damaging agents that produce multiple breaks in both single- and double-stranded DNA, with release of single nucleotides. This reaction requires the presence of Fe^{2+}, with which bleomycin forms a complex, and of molecular oxygen. Bleomycins seem to exert a quasi-catalytic action, as one molecule can repeatedly produce several breaks. These antibiotics are used only against some types of tumors.

2. *Inhibitors that form complexes with DNA.* There are many substances, both natural and synthetic, that form reversible complexes with DNA. The majority of such substances contain in their structure a planar polycyclic system, which is able to insert between the base pair of duplex DNA: for this reason they are called *intercalators.* Among the synthetic products, phenanthridines and acridines have been important, both in the laboratory as tools for studying genetic phenomena at the molecular level, and in clinics as antiprotozoal agents. Among the natural products, the following ones have been particularly studied.

- *Actinomycin D.* This is one of the first antibiotics to be isolated and is historically very important. Because of its toxicity, its clinical use is now limited to a few tumor forms, but it has been extremely useful in the study of transcription and replication. Actinomycin D complexes with DNA by intercalation of the tricyclic part of the molecule at some GpC sequences. The peptide rings of the antibiotic contribute to the stability of the complex by a lateral interaction with the DNA strands. Although actinomycin D inhibits the syntheses of both DNA and RNA, under special conditions the

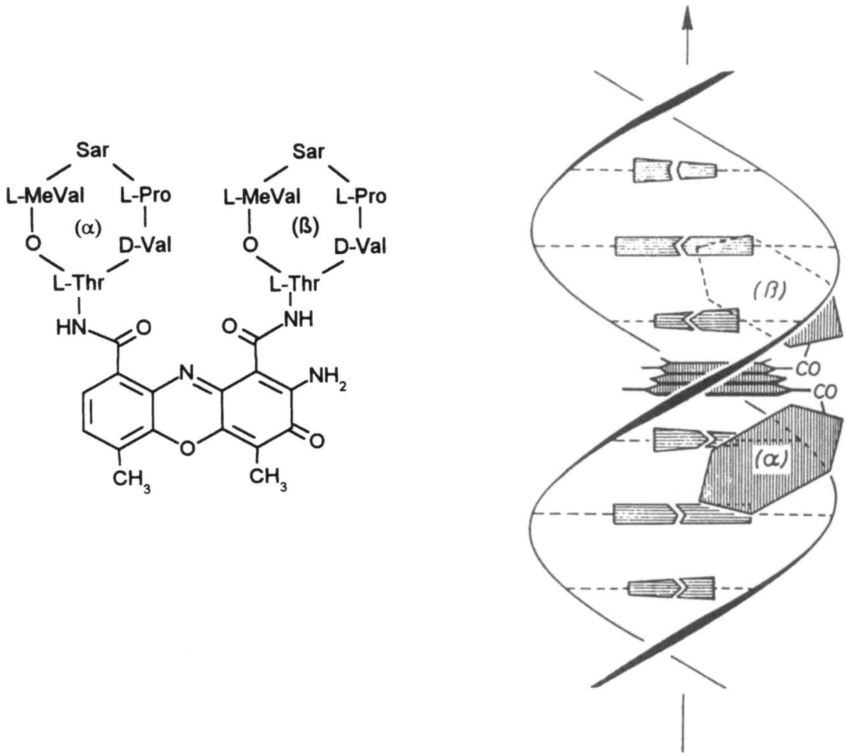

Figure 3.8. Chemical structure of actinomycin D and model of its interaction with DNA. [Redrawn from H. Lackner, *Angew. Chem. Int. Ed. Engl.* **14**:375 (1975).

latter appears more affected. It has been suggested that this effect is the result of the ability of the antibiotic peptide rings to mimic the termination signals of transcription and consequently to promote an early detachment of the nascent RNA chains (see Figure 3.8).

- *Daunorubicin* (*daunomycin*) and *doxorubicin* (*adriamycin*). These anthracyclines are among the most effective antitumor antibiotics. The molecule is composed of four hexatomic rings linearly condensed to give a planar structure bearing an amino sugar. The molecule can preferentially intercalate along the portions of DNA strands that are composed of alternating purines and pyrimidines. The effect of these anthracyclines is remarkably bactericidal and cytocidal. This latter effect is not understandable on the basis of a simple intercalation. The current interpretation is that the distortion caused by the antibiotic in the structure of the double helix prevents the completion of the reactions catalyzed by topoisomerase II. In fact, the enzyme performs the cleaving of the two strands, whose reactive ends bind to tyrosine residues of the enzyme. The distortion caused by the antibiotic prevents the second reaction of the topoisomerase from occurring, i.e., there is no restoration of the integrity of the strands after the unwinding of the chain, leaving multiple breaks along the DNA strands.

- *Distamycin* and *netropsin*. These products possess antiprotozoal and antitumor activity, and unlike those previously described, form complexes with DNA not by intercalation between the bases but by insertion into the minor groove of the double helix at specific sequences, particularly AATT.

3.4.2.2. Inhibitors of Replication Enzymes

No specific inhibitor of bacterial DNA polymerases has been reported to date. One inhibitor of eukaryotic DNA polymerase and a number of topoisomerase II inhibitors are known. The effects of a gyrase inhibitor (nalidixic acid) on macromolecular syntheses are illustrated in Figure 3.9, exemplifying inhibitors of replication enzymes.

Aphidicolin is a specific inhibitor of DNA polymerase α in eukaryotes

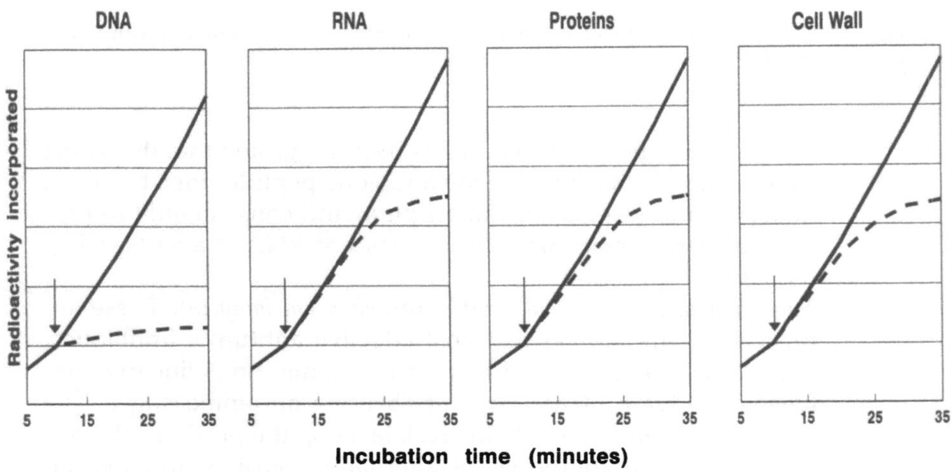

Figure 3.9. Effect of nalidixic acid on incorporation of radioactive precursors into DNA, RNA, proteins, and cell wall of growing bacteria. The solid line indicates incorporation into a control culture, the dashed line incorporation into a culture to which antibiotic was added at the time indicated by the arrow.

which has the same functions as DNA polymerase III in prokaryotes. The inhibition is of the allosteric type and is reversible; the effect is thus static and nonlethal.

Novobiocin and *coumermycins*. In the past, different mechanisms of action have been attributed to these antibiotics, as a consequence of the pleiotropic effects that are observed when DNA synthesis is inhibited. Today, it has been ascertained that the primary effect is the inhibition of DNA gyrase in bacteria. Novobiocin and coumermycins bind to the ATP site on the gyrase subunit B and prevent completion of the reactions catalyzed by the enzyme. At very high concentrations, they also inhibit topoisomerase II in eukaryotes.

Quinolone antibacterials. The best known among these synthetic products are: nalidixic acid, oxolinic acid, pipemidic acid, and the more recent group of 4-fluoroquinolones, such as norfloxacin and ciprofloxacin. These quinolones bind to the gyrase subunit A and therefore have a molecular site of action different from that of novobiocin and the coumermycins. This explains the absence of cross-resistance between these two classes of antibiotics, despite acting on the same target enzyme.

3.4.2.3. Inhibitors of Transcription Enzymes

Various antibiotics that specifically inhibit the bacterial RNA polymerase are known. Among these, *streptolidigin*, *lipiarmycin*, and *sorangicin* deserve to be mentioned, besides the *ansamycins*, which will be discussed in more detail. The common characteristics of their action are:

1. As prokaryotic and eukaryotic RNA polymerases are different, these antibiotics are usually selective and inhibit the growth of bacteria but not that of fungi or mammalian cells.
2. In growing bacteria, the synthesis of RNA is specifically inhibited without any direct effect on DNA synthesis. The block of messenger RNA synthesis causes, within a few minutes, cessation of protein synthesis (Figure 3.10).
3. As the temporary suspension of RNA synthesis is not in itself lethal, these antibiotics are bacteriostatic unless the complex formed with the enzyme is virtually irreversible.

Ansamycins. The structure of these molecules includes an aromatic group spanned by a bridge consisting of an aliphatic chain, called the "ansa." The ansamycins that contain a naphthalene ring and a 17-atom ansa (rifamycins and streptovaricins) are inhibitors of bacterial RNA polymerase. The other ansamycins show different mechanisms of action.

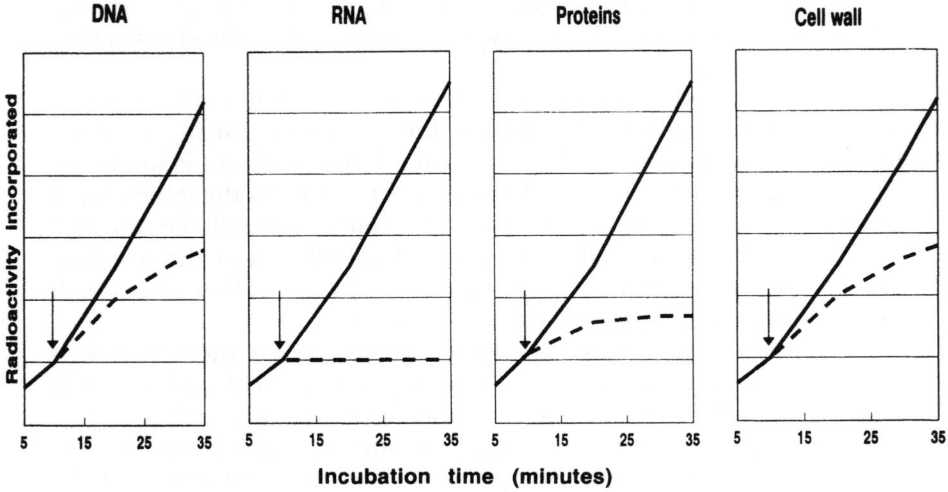

Figure 3.10. Effect of rifampin on incorporation of radioactive precursors into DNA, RNA, proteins, and cell wall of growing bacteria. The solid line indicates incorporation into a control culture, the dashed line incorporation into a culture to which antibiotic was added at the time indicated by the arrow.

Rifamycins are the only ansamycins produced commercially. In particular, rifampin, a semisynthetic derivative, is used extensively in the clinic as an antitubercular and antistaphylococcal agent. The action of rifamycins consists of the formation of a virtually irreversible complex with the RNA polymerase β subunit (see Section 3.2). Since formation of the complex does not take place when transcription of a chain has already initiated, it was hypothesized that these antibiotics may inhibit transcription initiation. However, it was later demonstrated that rifamycins do not inhibit formation either of the initiation complex between DNA and RNA polymerase or the first phosphodiester bond, but instead they inhibit formation of the second phosphodiester bond and thus elongation. It is worth mentioning that lipiarmycin, cited above, is a true inhibitor of RNA synthesis initiation.

3.5. Inhibitors of Protein Synthesis

3.5.1. Phases of Protein Synthesis

The genetic information is translated by the process of protein synthesis (*translation of the genetic information*), by which the amino acids are polymerized according to an order determined by the sequence of the nucleotide triplets in messenger RNAs. These nucleotide sequences are, in turn, determined by the sequence of deoxynucleotides in specific tracts of DNA.

Protein synthesis can be divided into two main processes: (1) activation and identification of the amino acids by means of attachment to transfer RNA (tRNA) and (2) polymerization of the amino acids, which takes place on the ribosomes and is known as the *ribosomal cycle*.

3.5.1.1. Activation and Identification of the Amino Acids

There is a specific enzyme for each amino acid, the aminoacyl-tRNA synthetase, that catalyzes the formation of an ester linkage between the carboxyl of the amino acid and a hydroxyl on the last nucleotide of its specific tRNA.

Energy for formation of the bond is provided by ATP. The amino acid is "activated" (the energy in the ester bond is sufficient for the subsequent formation of the peptide bond) and "identified" by being attached to the proper tRNA. This process is also called charging of the tRNA and the resulting molecule is called aminoacyl-tRNA. In bacteria there is a particular aminoacyl-tRNA, the methionyl-tRNA^{et}, which is formylated on the amino group and functions as the initiator of the polymerization process (tRNA initiator).

3.5.1.2. Ribosomal Cycle in Prokaryotes

Initiation Phase. The process begins with formation of a ternary complex between the 30 S subunit of the ribosome, which binds to a region of the messenger RNA containing the initiation triplet AUG, and the tRNA initiator, which is positioned on the same triplet. Three proteins, the *initiation factors* IF1, IF2, and IF3, contribute to the correct formation of the complex. The energy is provided by hydrolysis of GTP to GDP. The initiation triplet is recognized by the ribosome on the basis of specific sequences that are located 10–14 nucleotides upstream in the mRNA. The 50 S subunit then joins the complex to complete the ribosome. The tRNA initiator and the corresponding triplet are located in a site of the ribosome that is called site P (site of the nascent peptide chain) (Figure 3.11a).

Elongation Phase. A new aminoacyl-tRNA is positioned on the triplet next to the initiation triplet by the action of the *elongation factors* EF-Tu and EF-Ts in a site of the ribosome called site A (aminoacyl-tRNA

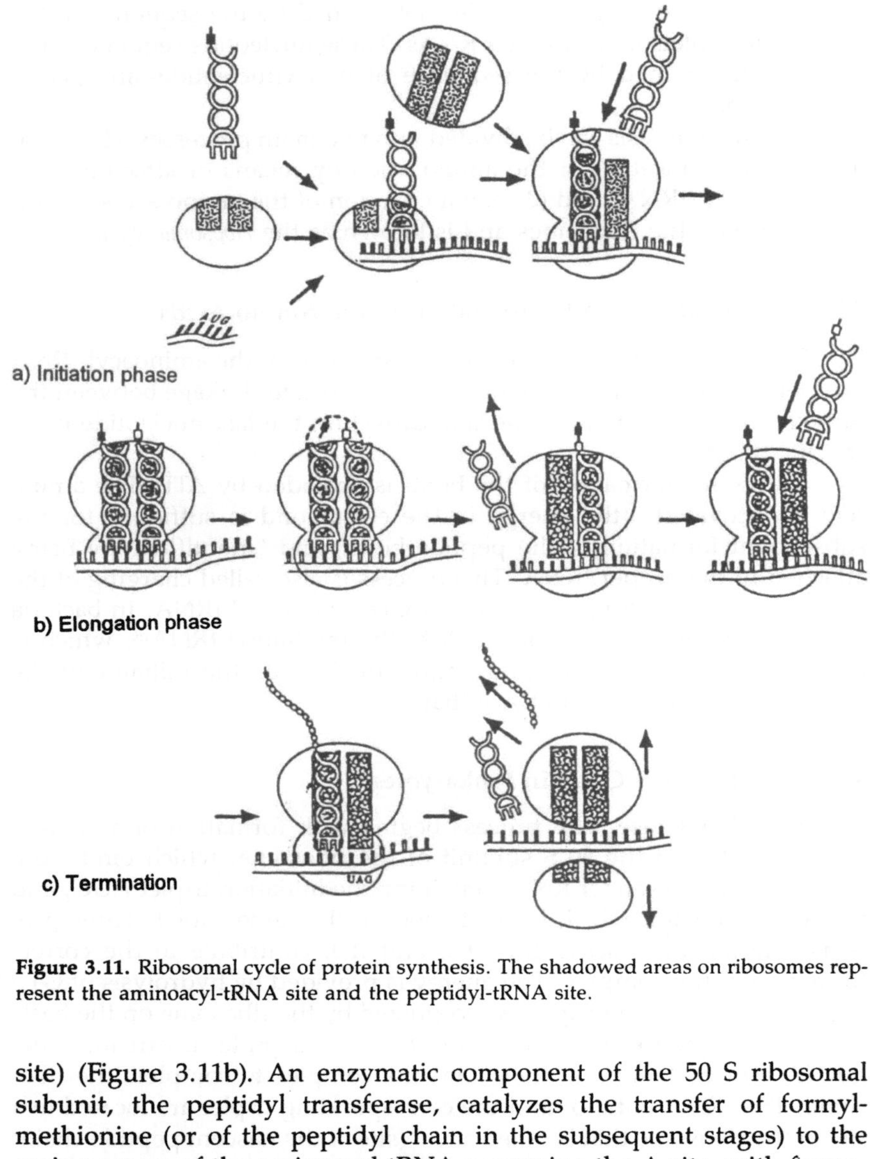

Figure 3.11. Ribosomal cycle of protein synthesis. The shadowed areas on ribosomes represent the aminoacyl-tRNA site and the peptidyl-tRNA site.

site) (Figure 3.11b). An enzymatic component of the 50 S ribosomal subunit, the peptidyl transferase, catalyzes the transfer of formylmethionine (or of the peptidyl chain in the subsequent stages) to the amino group of the aminoacyl-tRNA occupying the A site, with formation of a dipeptidyl-tRNA. By the action of another protein factor, EF-G, the ribosome is translocated by one codon (three bases) on the mRNA. The dipeptidyl-tRNA is thus shifted from the A site to the P site and the uncharged tRNA is released. The process is repeated and thus the polypeptide chain grows by adding one residue at a time to the carboxyl end of the chain. The energy for this series of reactions is provided by the hydrolysis of GTP to GDP.

Termination Phase. The process continues until the ribosome arrives at a specific triplet on the messenger (UAA, UAG, or UGA), which is the *termination signal* (Figure 3.11c). At this point the complex of peptidyl-tRNA, ribosome, and mRNA dissociates and the components reassemble to start the synthesis of a new chain. The process of termination is facilitated by specific proteins, the *termination factors* RF1, RF2, and RF3.

3.5.1.3. Ribosomal Cycle in Eukaryotes

The ribosomal cycle in eukaryotes is quite similar to that in prokaryotes. The main differences, besides the size and composition of ribosomes, are:

1. The initiator tRNA is not formylated.
2. The initiator factors, termed eIF1, eIF2, eIF3, and several eIF4, are different in their number and composition from those of prokaryotes.
3. The elongation factors EF-1α, EF-1β, and EF2 have the same functions as EF-Tu, EF-Ts, and EF-G, but again differ in composition. In yeasts an additional elongation factor is present, termed EF3.
4. Termination is assisted by a single protein factor, RF.

3.5.2. General Properties of Protein Synthesis Inhibitors

Inhibitors of protein synthesis act by different mechanisms and at different stages. Although their overall effects on the cell are not identical, some generalizations can be made:

1. Temporary arrest of protein synthesis is not lethal to the cell. In fact, it always occurs when, for example, a bacterial culture lacks essential nutrients. Therefore, inhibitors of protein synthesis are bacteriostatic, unless they form irreversible bonds with some essential component of the synthetic system.
2. The effects of blocking protein synthesis on the synthesis of other macromolecules are complex (e.g., chloramphenicol, Figure 3.12) and can be summarized as follows:

a. Replication of DNA chains already initiated continues to completion but there is no new chain initiation. In a growing culture, the overall effect is an apparent decreased rate of DNA synthesis.
b. RNA synthesis continues for a time, and may even in some cases be stimulated if the antibiotic acts at the ribosomal level; if, however, an antibiotic inhibits the charging of tRNA, in most bacteria there will be a block of RNA synthesis as an effect of the stringent response.
c. The rate of synthesis of cell wall declines and eventually ceases entirely.

3. The process of protein synthesis is more or less the same in all organisms. However, as previously described, some components do differ in different taxonomic groups. The principal differences between cells of prokaryotes and eukaryotes are listed in Table 3.2.

Protein synthesis inhibitors have selective effects when they inhibit structures or components that differ in the various groups of organisms, and nonselective effects when they inhibit components common to all cells. Therefore, there are selective inhibitors of protein synthesis in

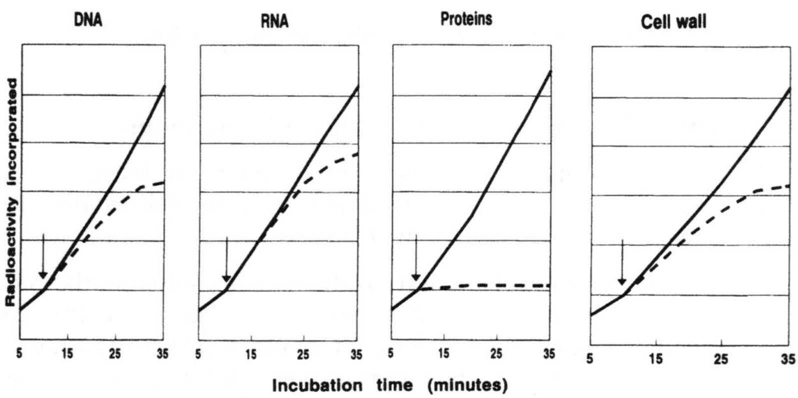

Figure 3.12. Effect of chloramphenicol on incorporation of radioactive precursors into DNA, RNA, proteins, and cell wall of growing bacteria. The solid line indicates incorporation into a control culture, the dashed line incorporation into a culture to which antibiotic was added at the time indicated by the arrow.

Table 3.2. Differences between Prokaryote and Eukaryote Protein Synthesis[a]

Components of protein synthesis mechanism	Eukaryotes	Prokaryotes
Ribosomal size	80 S	70 S
Ribosomal composition		
Proteins	60%	40%
RNA	40%	60%
Ribosomal RNA	5 S, 29 S, 18 Sa 5 S, 25 S, 16 Sb	5 S, 23 S, 16 S
Initiator aminoacyl-tRNA	met-tRNAi	F-met-tRNAi
Initiator factors	Different in the two groups	
Elongation factors	Different in the two groups	

[a] a = animals; b = plants; S = Svedberg unit, a measurement of the sedimentation coefficient. The sedimentation rate of a structure is a function of its size.

prokaryotes, selective inhibitors of protein synthesis in eukaryotes, and nonselective inhibitors. Antibiotics belonging to the last two classes are obviously toxic, unless there are substantial differences in permeation. One must also remember that, in eukaryotic cells, structures such as mitochondria have special protein-synthesizing systems that resemble those of bacteria. Antibiotics that act selectively on the prokaryotic system may, therefore, inhibit protein synthesis in these organelles and thereby cause side effects in eukaryotic cells.

With regard to ribosomal binding sites, antibiotics that interfere at site P are usually selective, whereas those that interfere at site A inhibit nonspecifically both eukaryotes and prokaryotes.

3.5.3. Examples of Inhibitors of Protein Synthesis

Antibiotics that inhibit protein synthesis are a large and diverse group and some of them have important clinical applications. They may be conveniently divided into three subgroups, according to their site of action:

1. Inhibitors of aminoacyl-tRNA formation
2. Inhibitors of ribosomal functions
3. Inhibitors of extraribosomal factors

3.5.3.1. Inhibitors of Aminoacyl-tRNA Formation

Only a few antibiotics are known that interfere with the synthesis of aminoacyl-tRNA. Among them, *borrelidin*, an inhibitor of threonine transfer, and *indolmycin*, which interferes with the transfer of tryptophan, do not have clinical applications.

Mupirocin acts on gram-positive bacteria by competitively inhibiting isoleucyl-tRNA-synthetase. Its action is bacteriostatic. Its lack of efficacy against gram-negative bacteria is related to its lipophilic nature, which prevents its permeation through the pores of the outer membrane. Although its action is selective and nontoxic, it is only used topically because it is rapidly metabolized into inactive products when administered systemically.

3.5.3.2. Inhibitors of Ribosomal Functions

Thermorubin has a broad spectrum of activity and is selectively active on bacterial protein synthesis. However, it is not used clinically as it is insoluble and inactivated by serum. Thermorubin specifically inhibits the initiation of protein synthesis, as demonstrated by the fact that *in vitro* it does not block the poly (U)-directed polyphenylalanine synthesis (a system representing elongation only) but does block correctly initiated systems directed by natural mRNA. In addition, it has been shown that thermorubin prevents the formation of the complex among mRNA, tRNA initiator, and ribosome, but that, once this complex is formed, it does not affect the subsequent reactions.

Kasugamycin is used as an antifungal in agriculture. It inhibits protein synthesis mainly at the initiation level both in prokaryotes and in eukaryotes. Inhibition is the result of the formation of a complex with the terminal part of the 16 S rRNA involved in the recognition of mRNA sequences; consequently, the initiator tRNA cannot bind to form the initiation ternary complex.

Streptomycin is a bactericidal antibiotic that binds irreversibly to a site on the 30 S subunit of ribosomes near the interface with the 50 S subunit. The resulting distortion of site A prevents the correct positioning of the aminoacyl-tRNAs. In particular, binding of methionyl-tRNA^{et} is hampered and transcription initiation prevented. *In vitro*, the distortion induces wrong couplings between the tRNA and the mRNA codon triplets, with the consequent formation of altered proteins. It has been claimed that the activity of streptomycin is related to the formation of a complex with the ribosomal protein S12. However, it was subsequently shown that streptomycin binds directly to specific bases of 16 S RNA and that the protein S12 contributes to its affinity for this site.

Gentamicin. The mechanism of action of gentamicin can be considered typical for the aminoglycosides containing deoxystreptamine, such as *tobramycin*, *kanamycin*, *amikacin*, and *sisomicin*. In contrast to streptomycin, these antibiotics bind to more than one ribosomal site, a fact reflected in the absence of one-step resistance. Again, highest affinity is displayed toward the 30 S subunit, but at a site different from that of streptomycin. The effect of these aminoglycosides is also different; mainly they inhibit translocation. These antibiotics are bactericidal probably because of a higher affinity for their target.

Tetracyclines are inhibitors of functions of both 70 and 80 S ribosomes, albeit to different extents. The effect is the result of the formation of a reversible complex between tetracyclines and a region of the 16 S rRNA near the site of attachment of the aminoacyl-tRNAs. Although this mechanism of action is nonselective, these antibiotics possess selective toxicity against bacteria, as there is a specific permeation by an active transport system in prokaryotic cells, whereas eukaryotic cells actively export these drugs.

Puromycin is used only in the laboratory as a "biochemical reagent." It can be viewed as a structural analog of the terminal 3′ end of aminoacyl-tRNA. As such, it binds to the acceptor site A of ribosomes of both prokaryotes and eukaryotes. It participates as an acceptor in the peptidyl transfer reaction, i.e., the terminal carboxyl of the growing peptide forms a peptide bond with the amino group of puromycin, which falls off the ribosome, thus terminating protein synthesis in the elongation phase.

Erythromycin. This is considered the representative antibiotic of the antibacterial macrolide family. It selectively inhibits prokaryotic protein synthesis by binding to the 50 S subunit. The binding site has been the subject of controversy, as experiments suggested that erythromycin could bind to various ribosomal proteins. The current hypothesis is that erythromycin binds directly to a specific region of the 23 S rRNA. The effect of the resulting structural distortion seems to be an early dissociation of the peptidyl-tRNA from ribosomes.

Lincomycin and *clindamycin* inhibit protein synthesis by binding to the 50 S subunit. Surprisingly, they complex with ribosomes of gram-positive bacteria but not with those of gram-negative origin. The binding site in the ribosome is at least partly coincident with that of erythromycin (also in this case the effect is a block of the peptidyltransferase function), with which they demonstrate partial cross-resistance.

Chloramphenicol acts on the 50 S subunit to inhibit the peptidyl transfer reaction. As recent experiments indicate that this reaction is directly mediated by ribosomal RNA, chloramphenicol is thought to bind to this macromolecule. However, various proteins actually contribute to the formation of the complex. The region of rRNA to which chloramphenicol binds is near, or functionally connected to, the binding region of erythromycin, as indicated by the partial cross-resistance between the two antibiotics.

Thiostrepton is frequently used in molecular biology as a selective marker in recombinant DNA studies. It binds with high affinity to rRNA of the 50 S subunit and inhibits the GTP-dependent reactions. The binding is facilitated by the L11 protein.

Cycloheximide is specifically active on ribosomes of eukaryotes and not on those of prokaryotes. For this reason, it is used for selectively inhibiting the growth of molds. It is believed to inhibit translocation mediated by the EF2 factor.

3.5.3.3. Inhibitors of Extraribosomal Factors

Fusidic acid is a steroidal compound that interferes with the function of the elongation factor EF-G. In its presence, the ternary EF-G–GDP–ribosome complex is stabilized and translocation of ribosomes is inhibited. The EF-G factor is "frozen" in the ternary complex and, therefore, it is no longer available for another translocation cycle. *In vitro*, fusidic acid has an inhibitory action on the EF-2 factor of eukaryotes. Its selectivity of action is probably related to selective permeation.

Elfamycins (*kirromycins*). The name of this group of antibiotics relates to their mechanism of action, the specific inhibition of the elongation factor EF-Tu (Elfa = elongation factor). Kirromycins induce an allosteric alteration of the enzyme structure and interfere with the reactions of the aminoacyl-tRNA–EF-Tu–GTP complex. It is noteworthy that kirromycins possess an uncommon selectivity of action, as they are active only against some species of gram-positive bacteria, such as streptococci, and inactive against staphylococci. This seems to be related to differences in the structure of the factor EF-Tu. Elfamycins also display inhibitory activity when tested in *in vitro* systems utilizing gram-negative bacteria, but are inactive against intact cells, because of lack of penetration.

Antibiotic GE2270. This is a product isolated from cultures of *Planobispora rosea*, and is presently under development. Although it belongs to the class of thiazolylpeptides, such as thiostrepton and nosiheptide, it differs from these in its mechanism of action. In fact, like elfamycins, it binds to and inhibits the elongation factor EF-Tu. However, the binding site is different from that of elfamycins, which is reflected in the absence of cross-resistance and in susceptibility differences among the various bacterial species.

3.6. Inhibitors of Cell Membrane Functions

Both prokaryotic and eukaryotic cells are surrounded by a cell membrane (or plasma membrane) that controls the bidirectional flow of sub-

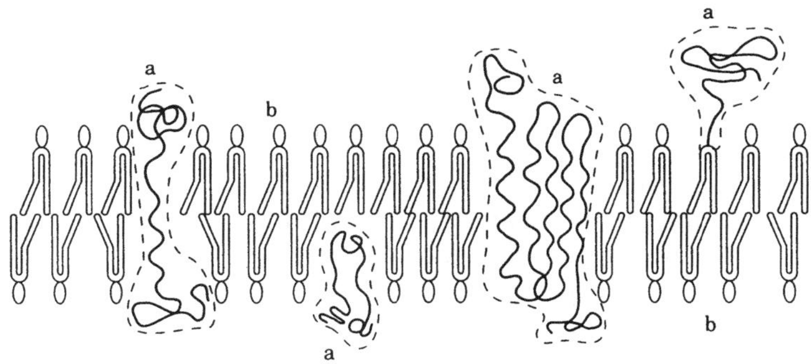

Figure 3.13. Schematic representation of membrane structure according to the classic "mosaic" model. a, proteins; b, lipids.

stances between the outside and the cell interior. The supermolecular structure of the cell membrane consists essentially of a double lipid layer in which protein molecules are intercalated, as in a mosaic (Figure 3.13). The basic structure of the membrane is preserved throughout the phylogenetic ladder from bacteria to mammalian cells, with an important difference: there are no sterols in bacterial cell membranes, ergosterol is present mainly in cell membranes of fungi, and cholesterol in those of mammals.

Antibiotics that act on the cell membrane can be divided into two groups:

1. Substances that disorganize the supermolecular structure of the membrane, thus causing loss of cellular components
2. Compounds that act as carriers for specific ions (for this reason they are called *ionophores*) and cause either an abnormal accumulation or an abnormal excretion of ions

Generally, these antibiotics are poorly selective, acting against both bacterial and eukaryotic cells, reflecting the considerable chemical and structural similarities of the cell membranes of different organisms. As a consequence of this lack of specificity, they are usually too toxic to be given systemically and their use is often limited to topical applications.

3.6.1. Examples of Inhibitors Affecting Cell Membrane Functions

3.6.1.1. Substances That Disrupt the Membrane Structure

Polymyxins are a family of antibiotics characterized by a heptapeptide ring bearing basic hydrophilic groups and a lipophilic chain. The hydrophilic groups complex with the phosphoric groups of phospholipids, while the aliphatic chain can insert itself between the lipidic chains, thus disaggregating the membrane structure. Polymyxins are more active against gram-negative than gram-positive bacteria, probably because they have a higher affinity for the outer membrane than for the cytoplasmic membrane.

Polyene antibiotics. The characteristic structure of these substances consists of an aliphatic chain closed into a ring by a lactone bond. The chain includes a lipophilic region comprising a series of conjugated double bonds and a hydrophilic region bearing a series of hydroxyl groups. The lipophilic region has a great affinity for membrane sterols. The kind of cytolytic alterations, e.g., pores, caused by the formation of complexes between polyene antibiotics and the membrane sterols are still a matter for investigation. It has been suggested that pores allowing the passage of ions and other water-soluble molecules are formed, lined with the hydrophilic region of polyenes. A current hypothesis is that some polyenes interact with greater affinity with ergosterol (the sterol of fungal membranes) than with cholesterol (the sterol of animal cell membranes) and therefore can be used as antifungal agents. The two most important antibiotics of this family are *amphotericin B* and *nystatin*.

3.6.1.2. Ionophores

These antibiotics are a large and diverse group of molecules with the ability either to form lipid-soluble complexes with cations or to form pores through which cations can cross the lipophilic cell membrane. The consequence can be:

1. An accumulation of cations in the cell
2. An abnormal loss of cations
3. Sequestration of cations present in the environment making them unavailable to the cell

The different ionophores differ in the molecular mechanisms by which they act and in their specificity toward different cations. In terms of the mechanism of action these antibiotics can be grouped as follows:

1. *Stationary ion-conducting channels*. These are molecules able to insert themselves into the structure of the cell membrane, forming pores whose walls are lipophilic on the outside and hydrophilic on the inside, thus inducing cation leakage from the cell (Figure 3.14). These antibiotics lack prokaryote–eukaryote specificity and cannot be used as systemic antibacterial drugs. Members of this group include the linear peptides *gramicidins A, B,* and *C* and the cyclic peptide *alamethacin*. Gramicidins, which are active against certain gram-positive bacteria, are used topically in some countries.

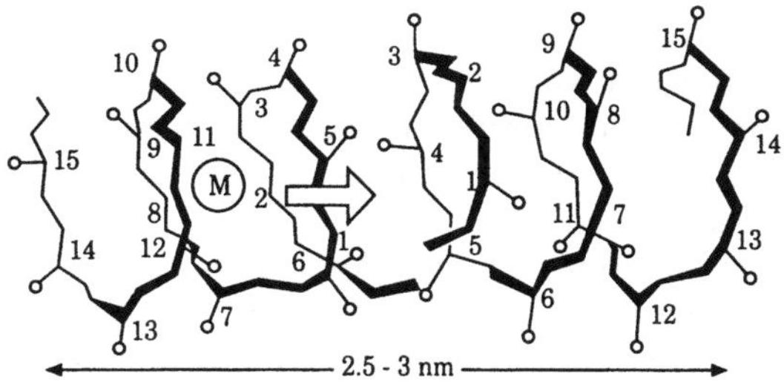

Figure 3.14. Schematic representation of the arrangement of two gramicidin A molecules to form a pore through the cell membrane. Numbers refer to the 15 amino acids of the polypeptide; M represents an ionic metal. [Redrawn from Y. Ovchinnikov, *Eur. J. Biochem.* **94**:321 (1979).]

2. *Mobile ionophores.* These are molecules that can complex with ions and are able to cross the cellular membrane, thus inducing an abnormal transcellular distribution of specific cations. Chemically, they belong to three major classes.

a. Cyclodepsipeptides. The typical representative of this class is *valinomycin,* whose ring is formed by six amino acids and six hydroxyacids regularly alternating. Noteworthy is the stereochemical configuration of the acidic units, which are in alternating couples D,D and L,L. The resulting spatial structure is a fairly rigid ring, with the ester carbonyls forming an inner sphere that can specifically accommodate K^+ ions (Figure 3.15). Other cyclo-

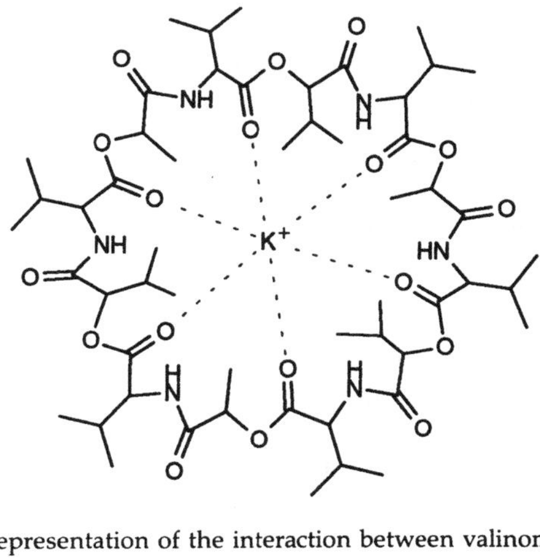

Figure 3.15. Representation of the interaction between valinomycin and K^+.

depsipeptides, e.g., *enniantins*, possess a smaller ring and therefore can complex Na^+ ions preferentially. The antibacterial action of valinomycin is attributable to depletion of K^+ in the cell and to the consequent interruption of the potassium-dependent enzymatic reactions.

b. Polyethers are linear molecules containing tetrahydrofuran or tetrahydropyran rings, whose ethereal oxygen atoms are arranged in such a way as to allow complexation with K^+ and Na^+ ions. The typical representative of this group is *monensin*, widely used as a coccidiostat and as an animalavian growth promoter. Other antibiotics of this group are *lasalocid* and *salinomycin*.
c. Sideromycins. These are antibiotics that complex ferric ions. *Deferoxamine* (desferrioxamine) is used to treat cases of toxicity related to iron overload. The antibacterial activity of sideromycins was believed to result from their ability to deprive the bacterial cell of ferric ions. However, it has been demonstrated, at least for *albomycin*, that the antibiotic moiety that binds iron actually functions as a carrier of the antibiotic into the cell, where this moiety is hydrolyzed releasing the active part of the drug.

3.7. Antimetabolites

The term *antimetabolites* refers to a group of natural and synthetic substances, with very heterogeneous chemical structures and mechanisms of action, but having in common the fact that their inhibitory effect can be antagonized by one or more metabolites. Generally, but not always, their chemical structures are analogous to those of the metabolites they antagonize.

Antimetabolites can be divided into two large groups on the basis of their mechanisms of action:

1. Those that are incorporated into "informational" polymers (DNA, RNA, proteins) in place of natural monomers, resulting in alteration of the information content
2. Those that inhibit the formation of essential metabolites

In both cases, antimetabolites generally have structures similar to those of natural metabolites of which they are analogues, and interact with the enzyme site that normally recognizes the natural ligand. The degree of inhibition depends on the relative affinity for the enzyme of the analog and the natural metabolite and on the ratio of their concentrations (*competitive inhibition*).

The effect of the action of compounds belonging to the second group can be overcome not only by the natural product to which they are analogous, but also by addition to the system of metabolites downstream in the metabolic pathway inhibited.

3.7.1. Examples of Antimetabolites Incorporated into Macromolecules

The polymerization system for amino acids sometimes cannot distinguish between an amino acid and one of its analogues and incorporates either into the polypeptide chain, producing abnormal proteins that have little or no function, and resulting in cell growth inhibition. The best known amino acid analogues acting in this way are *fluorophenylalanine* (a synthetic product) and *selenomethionine* (which occurs naturally in some plants).

Similarly, some analogues of nucleosides can be incorporated into DNA in place of the natural monomers. There are no important antibacterial antibiotics of this type. *Arabinosyl-cytidine* (*ara-C, cytarabine*), a synthetic analog of deoxycytosine, is used as an antitumor agent. *Vidarabine* (*arabinosyl-adenine, ara-A*) is a synthetic analog of deoxyadenosine, but has also been isolated from cultures of *Streptomyces antibioticus*. It has a respectable antiviral activity especially against herpesviruses. Both ara-C and ara-A are active only after metabolic phosphorylation, the active form being the corresponding nucleotides. Analogs of nucleosides rather than of nucleotides are used as the latter are unable to cross the cytoplasmic membrane. The efficacy of vidarabine is limited by its rapid deamination by the enzyme adenosine deaminase. To decrease this inactivation, one can combine vidarabine with *deoxycoformycin*, an antibiotic analogous to deoxyadenosine, which inhibits adenosine deaminase.

Iododeoxyuridine is another molecule incorporated into DNA after phosphorylation and is used to treat infections caused by herpesvirus. Its selectivity is related to the presence of higher levels of thymidine kinase, the phosphorylating enzyme, in the infected than in normal cells.

An important analog of thymidine is *azido-dideoxythymidine* (*AZT*), which is a selective inhibitor of retrovirus reverse transcriptase. It is used in the treatment of AIDS, a disease caused by HIV virus. In this case also, activity of AZT is dependent on its conversion into the triphosphate derivative. It has not been ascertained whether the inhibition is related to incorporation into the nucleic acids or to other mechanisms.

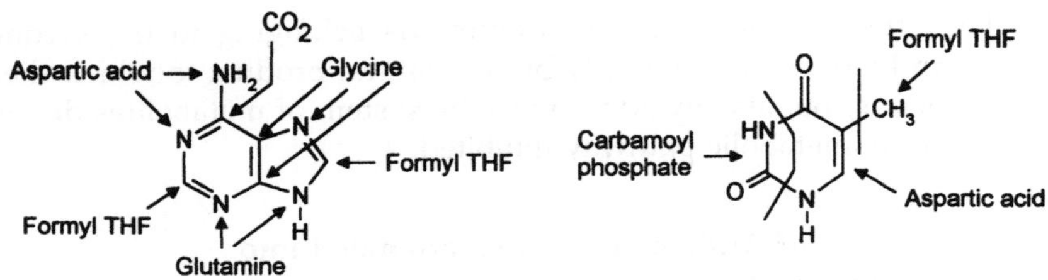

Figure 3.16. Metabolic origin of the carbon and nitrogen atoms in adenine and thymine.

3.7.2. Inhibitors of the Synthesis of Metabolites or Essential Cofactors

Unlike the substances described in the preceding section, many antimetabolites are not incorporated into the macromolecules in place of the metabolites they resemble, but inhibit one or more enzymes in the synthetic pathways of intermediate metabolism. Among the numerous compounds described in the literature, the most important are the inhibitors of nucleotide synthesis and the inhibitors of the synthesis and functions of folic acid. The origins of the carbon and nitrogen atoms in purine and pyrimidine bases are shown in schematic form in Figure 3.16. Many nitrogen atoms derive from aspartic acid or glutamine. It is not surprising that analogues of these amino acids inhibit the synthesis of these bases. Some carbon atoms derive from formyltetrahydrofolic acid: the products that interfere with its synthesis will result in the arrest of nucleotide synthesis.

As temporary depletion of nucleotides is not lethal, the action of these inhibitors is in general bacteriostatic. However, there is an exception when the only nucleoside missing is thymidine (leading to thymineless death). There are two theories to explain this fact: (1) synthesis of new DNA is initiated but, because of the absence of thymidine, only short chains are formed, which are exposed to attack by DNases; (2) in the absence of thymidine, uridine is incorporated into the DNA in its place, thus activating repair enzymes whose action may be lethal.

3.7.2.1. Inhibitors of Nucleotide Synthesis

Alanosine and *hadacidin*. These are inhibitors of purine biosynthesis and are structurally related to aspartic acid (Figure 3.17). Functionally they inhibit two different reactions in which aspartic acid acts as nitrogen donor: alanosine inhibits an early stage of ring formation, i.e., the

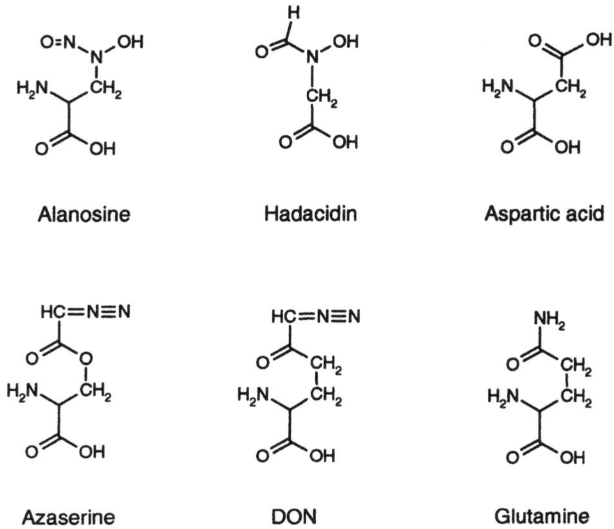

Figure 3.17. Antibiotics structurally analogous to aspartic acid and to glutamine.

introduction of nitrogen in position 1; hadacidin inhibits the introduction of the amine at C = 6, i.e., the conversion of inosine into adenosine.

Azaserine and *DON* are analogues of glutamine (Figure 3.17) and inhibit one of the initial reactions in purine synthesis, i.e., the introduction of nitrogen in position 3. Azaserine covalently binds to an SH group of the active site of the enzyme. DON acts probably in the same way but, in addition, inhibits the transfer of the amino group from glutamine to UTP in CTP synthesis.

Fluorouracil and *floxuridine*. These synthetic antitumor agents are metabolically converted into fluorodeoxyuridylic acid, a potent inhibitor of thymidylate synthetase, the enzyme that converts dUMP into dTMP. The effect at the cellular level is lethal, the result of "thymineless death."

3.7.2.2. Inhibitors of the Synthesis or Functions of Folic Acid

Although in this book only natural antibiotics are treated, it is worthwhile to briefly mention *synthetic* antibacterial agents that have interesting mechanisms of action.

Sulfonamides are structurally similar to *p*-aminobenzoic acid and compete with it in the synthesis of folic acid (Figure 3.18). They both exert an inhibitory action on synthesis and are incorporated, giving rise

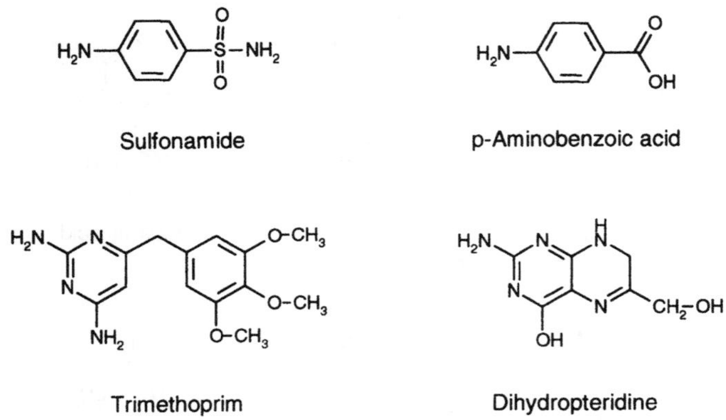

Figure 3.18. Comparison between the structures of sulfonamide and trimethoprim and the structures of *p*-aminobenzoic acid and dihydropteridine (components of folic acid), respectively.

to inactive analogues or direct inhibitors. In bacterial cultures, sulfonamide action is not antagonized by addition of folic acid because bacteria are impermeable to this coenzyme. On the other hand, the action is antagonized by methionine and by thymine, whose syntheses require folic acid. The sulfonamides are inactive in the cells of higher organisms as these do not have the ability to synthesize folic acid but obtain it from the environment.

Diaminopyrimidines. These are inhibitors of the enzyme dihydrofolate reductase that converts dihydrofolic acid into tetrahydrofolic acid, coenzymes to which they are structural analogues (Figure 3.18). Tetrahydrofolic acid is a cofactor in one-carbon-atom metabolism. As the structure of dihydrofolate reductase is different in different organisms, there are specific inhibitors of the enzyme of mammals, such as *methotrexate*, of protozoa, such as *pyrimethamine*, and of bacteria, such as *trimethoprim*. The effect of the latter depends on the presence or absence of thymidine, methionine, and glycine in the culture medium. If the three nutrients are present, there is no inhibition (because of reversion by end products). If none is present, the effect is bacteriostatic. If thymidine is absent but glycine and methionine are present, the effect is bactericidal (a typical case of thymineless death, mentioned above).

3.8. Selectivity of Antibiotic Action

The study of antibiotics is essentially an applied science. It is directed at discovering and using substances that can inhibit the growth of microorganisms parasitic to man or to higher animals without damaging the host organism. Therefore, an antibiotic must be selective in its mechanism of action, i.e., it must damage the bacterial and not the host cells (*selective toxicity*).

3.8.1. Evolutionary Basis of Selectivity

The selectivity of antibiotic action derives from the structural and biochemical variations that have arisen during evolution without changing the essential functions of the cell.

On the evolutionary scale, one can distinguish between two principal groups of organisms: prokaryotes and eukaryotes. The prokaryotes include all the bacteria and blue algae. All the other cellular organisms (yeasts, fungi, algae, protozoa, plants, and animals) are eukaryotes. There is also a third group, the Archaea (formerly Archaeobacteria), having a general structure similar to that of prokaryotes but with remarkable biochemical differences, e.g., the composition of the cell membrane and wall. These are not discussed because they are neither pathogenic nor producers of antibiotics. The most important structural and biochemical differences between prokaryotes and eukaryotes are summarized in Table 3.3.

3.8.2. Cellular and Molecular Basis of Prokaryote versus Eukaryote Selectivity

Why does a given concentration of an antibiotic inhibit the growth of a given microbe but remain practically inactive against another microbial species? Why does an antibiotic that effectively inhibits bacterial cells at low concentrations remain inactive even at much higher concentrations against animal cells? In other words, what are the cellular and molecular bases of the selectivity of antibiotic action?

Antibiotics are a very heterogeneous group of substances, both in terms of structure and mechanism of action, and thus one should expect that their specificities would have different bases. The selectivity of antibiotic action can be related to one or more of the following reasons:

1. Differences in the degree of permeation between different organisms
2. Action on essential structures or metabolic processes present in microorganisms but absent in animal cells
3. Different affinities of the antibiotic for enzymes that, while playing the same role in different types of cells, are structurally different

Table 3.3. Synopsis of the Essential Cellular Differences between Eukaryotes and Prokaryotes, Responsible for the Selective Toxicity of Antibiotics

Natural structures	Eukaryotes	Prokaryotes
Nucleus	True nucleus enclosed by a membrane	Absence of a nuclear membrane
Chromosomes	Several chromosomes made up of DNA and proteins (histones)	One chromosome made up of DNA only
Mitotic apparatus	Presence of centriole and spindle	Absence of mitotic apparatus; role played by plasma membrane
Cytoplasmic membrane	Presence of sterols	Absence of sterols
Cell wall	Absent in animals. When present (algae, plants, and fungi) is made up of simple polysaccharides	Typically a complex structure with peptidoglycans and atypical amino acids
Ribosomes	80 S, except for those of mitochondria and of chloroplasts (70 S)	Only 70 S
Respiratory systems	Located in specialized organelles: mitochondria	Located on plasma membrane
Cytoplasmic organelles	Present (mitochondria, chloroplasts)	Absent
Cellular osmotic pressure	Similar to that of environment	Higher than that of environment

3.8.2.1. Differences in Permeability

Except for those acting on the cell surface, antibiotics must enter the cell and reach a concentration sufficient to inhibit cell growth. The intracellular concentration of an antibiotic at any one moment is dependent on the rate of penetration and the rate at which it becomes diluted as a result of cell division, and, in some cases, of excretion. Differences in permeability occur mainly for substances that cross the cytoplasmic membrane by specific transport systems, which can differ in different organisms.

Tetracyclines are an example of selective permeability. These antibiotics inhibit protein synthesis in cell-free systems both from bacteria and from animal tissues. However, they inhibit the growth of bacterial cells but not that of animal cells, as they penetrate into bacteria by an efficient and specific transport system whereas their internal concentrations in animal cells never reach inhibitory levels.

3.8.2.2. Essential Structures Present in Bacteria but Not in Higher Organisms or Vice Versa

An essential difference between microorganisms and cells of higher organisms (especially animals) is the presence in microbial cells of a cell wall, a rigid envelope that prevents the cell from bursting as a consequence of its high internal osmotic pressure. The chemical composition and the physical structure of the cell wall are completely different in bacteria and in fungi (see Section 3.3). Some antibiotics specifically inhibit the synthesis of the bacterial cell wall and they will obviously be inactive against fungi, yeasts, and animal cells. Also, the inhibitors of cell wall synthesis in fungi or yeasts display, at least theoretically, a selective action.

Cell membranes are substantially similar in structure and function throughout the evolutionary scale from bacteria to mammals. However, there are small variations among the different groups. For example, the eukaryotic membrane contains various kinds of sterols, which vary in different organisms and which are not present in bacterial membranes. Certain antibiotics interfere with sterols of the eukaryotic cell membrane to an extent that causes irreversible damage. Since bacteria do not possess these sterols in their cell membranes, they are insensitive to these agents. As far as the metabolic processes are concerned, there are several differences between microorganisms and higher organisms. In particular, many amino acids and coenzymes are synthesized only by microorganisms and plants and, consequently, antibiotics that inhibit these biosynthetic processes will display a selective action.

3.8.2.3. Difference Affinity for Enzymes

There are many enzymes that, although playing similar roles in cell metabolism, are structurally quite different in prokaryotes and eukaryotes. Typical examples are RNA polymerases, of which only one if present in bacteria and three in eukaryotes, and elongation factors of protein synthesis, which are different in bacteria, fungi, and mammalian cells. Due to these different structures, several antibiotics have a high affinity for the enzymes of one class of organisms but do not bind to the others.

3.8.3. Cellular and Molecular Basis for Differences in Antibacterial Spectrum

Antibiotic activities show differences not only when comparing prokaryotes and eukaryotes but also among different species of prokaryotes. In other words, they have different antibacterial spectra. The reasons for the differences in the antimicrobial spectra of different antibiotics fall into three classes:

1. Differences in permeability
2. Presence or absence of inactivating enzymes
3. Differences in the cellular structure or in the enzymes with which the antibiotic interacts

3.8.3.1. Differences in Permeability

To enter the bacterial cell, substances must cross a barrier consisting of the cellular membrane and the cell wall. Whereas the membrane structure is substantially the same in all bacteria (with the exception of Archaea), that of the cell wall shows marked differences. From this viewpoint, bacteria can be divided into three major groups: *gram-positives, gram-negatives*, and the so-called *"acid fast,"* which includes *mycobacteria*. The structure of the walls has been described in Section 3.3. The gram-positive wall is easily permeable by substances such as antibiotics having a relatively low molecular weight, and, therefore, it does not act as a true permeation barrier. On the contrary, in the gram-negative wall the outer membrane acts as a true lipophilic barrier and can be crossed via passive transport only by substances that possess appropriate physicochemical characteristics, e.g., a certain degree of lipophilicity. However, some proteins are present in this membrane (*porins*), which promote the formation of channels, or *pores*, through which hydrophilic molecules of molecular weight of 600 or less can pass. The various porins show poor selectivity toward most of the solutes but may be truly selective toward anions and cations. The existence of the outer membrane is considered the major reason for the selective activity of antibiotics against gram-positive and gram-negative bacteria.

Molecular diffusion through the cellular membrane may occur by two principal mechanisms:

1. A passive transport mechanism, which depends on the concentration gradient between the outside and the inside of the cell, and on the physicochemical nature of the molecule
2. An active transport mechanism, which involves specific proteins as carriers and requires energy. The active transport of antibiotics into bacteria is generally the result of an "illicit" transport, i.e., the antibiotic structure is sufficiently similar to that of a nutrient to use up the natural transport system. Since transport mechanisms may differ in different bacterial species the susceptibility to the antibiotic may also be different.

3.8.3.2. Presence of Inactivating Enzymes

Several bacterial species possess the ability to produce enzymes that inactivate some antibiotics by modifying their chemical structure. This property is particularly evident in species of the genus *Pseudomonas*, thereby explaining (in addition to their impermeability) their lack of susceptibility to the majority of known antibiotics.

3.8.3.3. Selectivity Related to Cell Structure or Target Enzyme Differences

Although most antibacterial antibiotics are able to act on structures present in all bacterial species, there are some examples of antibiotic inactivity that are the result of a lack of appropriate target. A clear case is that of mycoplasmas, bacteria devoid of a cell wall and thus insusceptible to penicillins and other antibiotics acting on peptidoglycan synthesis. Another example is bicyclomycin, which is thought to be inactive against gram-positive bacteria as it interferes with the termination factor ρ, typical of gram-negatives.

There are also cases of selectivity related to structural differences in the enzymes that have the same function in different bacterial species. The typical example is that of the elfamycins (kirromycins), for which it has been demonstrated that the activity against the different gram-positive species depends on structural variations in their target, the EF-Tu factor (see Section 3.5.3.3).

Chapter 4

Resistance of Microorganisms to Antibiotics

This chapter is composed of three independent but integrated parts. In the first one we describe the modifications present in the *resistant bacterial cell* with respect to the corresponding *susceptible bacterial cell*. This topic is closely correlated to the chapter on mechanism of action. In the second part we describe the possible localizations and modes of transmission of the genetic determinants of resistance. The approach applied is a systematic presentation of the different types of resistance mechanisms.

Finally, the biochemical and genetic aspects of resistance are presented in some detail for three important classes of antibiotics: β-lactams, macrolides, and aminoglycosides, as in these three classes one can find examples of almost all types of resistance mechanism.

Chapters 3 and 4 are strictly integrated and, together, can be read as an autonomous section of the book.

4.1. General Aspects

It is known that many bacterial infections that until a few years ago could be cured with specific antibiotics are today *resistant* to them. This is a general phenomenon, as it has been observed in different parts of the world, at different times, and for numerous (almost all) antibiotics. Its clinical relevance is obvious. Bacterial resistance to antibiotics has been very actively investigated, determined and measured at the epidemiological level. In this chapter we concentrate on the biochemical and biological aspects of resistance. More precisely, we attempt to answer the following questions:

1. What has been changed in the resistant bacteria with respect to the sensitive (susceptible) ones?
2. What role does the antibiotic play in the appearance and dissemination of resistance?
3. Is resistance a genetic phenomenon, i.e., is the information for resistance contained within the bacterial DNA?
4. If so, is resistance transferred from one bacterial strain to another? How?

Before answering these questions, it is necessary to provide some definitions and a brief description of the methods used to measure resistance.

4.1.1. Definitions

Resistance. A bacterial strain derived from a species that is susceptible to an antibiotic is said to be resistant when it is not inhibited by the minimal concentration of the antibiotic that inhibits the growth of the typical strains of that species. This strain may of course be inhibited by a higher concentration of the same antibiotic. Therefore, it is obvious that the concept of resistance applies to

1. A given bacterial strain
2. A given antibiotic, and
3. A given concentration. In the clinical literature, bacterial resistance is often discussed without mentioning the concentration, but this omission is only apparent as reference is tacitly made to the antibiotic concentrations in serum or tissues. In fact, a bacterial strain is considered resistant to a given antibiotic if it can grow in the presence of a concentration equal to or greater than that which the antibiotic can reach in serum or tissues.

The methods employed to determine the resistance or the susceptibility of a bacterial strain toward one or more antibiotics are described in Chapter 2.

Multiple resistance. When a microorganism is resistant to two or more antibiotics through unrelated mechanisms of resistance, one speaks of *multiple resistance*.

Cross-resistance. There are available in the market, and even more so among the cell physiologist's tools, many antibiotics that inhibit bacterial growth by the same mechanism of action. Usually, a bacterial strain that is resistant to an antibiotic is also resistant to other antibiotics of the same family. This phenomenon, called *cross-resistance* or co-resistance, is of considerable practical importance. In fact, if an infection is the result of a strain resistant to a certain antibiotic, the presence or absence of cross-resistance is a key determinant in choosing an alternative suitable antibiotic. When resistance is related to the production of inactivating enzymes, the different members of a given class may present quite different susceptibilities toward inactivation.

One-step and multistep resistance. It is possible to isolate resistant mutants from a population of bacteria susceptible to a given antibiotic by seeding a large number of cells on a solid culture medium containing a concentration of antibiotic several times higher than the MIC. The great majority of cells will be inhibited, but if a sufficiently large number of cells, for example 10^9 or more, are used as inoculum, one may find resistant cells that give rise to easily identifiable colonies. Frequently, the bacteria in these colonies are resistant not only to the concentration of antibiotic used for the selection but also to higher concentrations. This is called *one-step resistance*, and occurs in the case of certain antibiotics for which the resistance is an all-or-none phenomenon, i.e., the cells are either susceptible to low concentrations or are not inhibited even by high concentrations. A typical example is resistance to streptomycin.

In the case of other antibiotics, e.g., penicillin, it is not possible to isolate resistant cells by the method described above. Usually, bacterial populations contain mutants resistant to the MIC of these antibiotics, but sensitive to slightly higher concentrations. Therefore, it is extremely difficult to find a concentration of antibiotic in solid medium that will inhibit the growth of all the normal cells and not of the resistant ones. Resistance of this type can be selected from liquid culture media by a method inappropriately called *training*. According to this method, the bacterial strain is grown in concentrations of antibiotic slightly below the MIC to select the less susceptible cells. These then give a population that includes cells capable of growing in the presence of slightly higher concentrations of antibiotic. By repeated passages of this type and by selecting from large populations in liquid culture, one can obtain strains that are notably more resistant. This gradual acquisition of resistance, found for penicillin and several other antibiotics, has been termed *multistep resistance*.

Resistance and insusceptibility. For the purpose of clarity, it is preferable to reserve the word *resistant* for those bacterial strains that derive from a susceptible species and have lost their susceptibility to a given antibiotic through mutation or other alteration of their genetic heritage. Bacterial species that are "constitutionally" not inhibited by an antibiotic because, for example, they lack the structure on which the antibiotic acts, are called *insensitive* or *insusceptible*. The expression *intrinsic resistance*, sometimes used instead of *insusceptibility*, should be avoided as it may create confusion.

There can also be sensitive bacteria that are not inhibited by an antibiotic because of environmental and culture conditions. For example, the MIC of streptomycin against *Staphylococcus aureus* is much lower at pH 8 than at pH 5. Under acidic conditions *S. aureus* will therefore be practically insensitive to streptomycin, although it is not resistant in the sense of the strict definition given above. In discussing these cases, some authors use the term *phenotypic resistance* because the phenomenon is not related to any change in the genetic heritage of the bacterial population.

Frequency of mutants and frequency of mutation. Frequency of mutants refers to the fraction of resistant cells present in a sensitive population. Frequency of mutation is the fraction of resistant mutants that originate in a population at each generation (each doubling).

The frequency of mutants differs numerically from the frequency of mutation for many reasons:

1. *The number of mutants present in a population derives both from those originating each generation and from the offsprings of the mutants formed in the preceding generations*; consequently, the frequency of mutants tends to increase generation by generation.
2. *Retromutation may occur generating susceptible cells from resistant ones, thus reducing the number of mutants.*
3. *The rate of growth of the resistant mutants may be different from that of the susceptible population*; if, as often is the case, the former is lower, the frequency of resistant mutants tends to decrease with further generations.

Determination of the frequency of mutants is relatively simple in the case of one-step resistance. In fact, it is sufficient to inoculate some plates of an agar culture containing the chosen concentration of antibiotic with a sufficiently large number of cells and, after incubation, to count the grown colonies. Their number divided by the number of cells inoculated represents the frequency of resistant mutants in that population.

The determination of the frequency of mutation is complex and is based on the fluctuation test of Luria and Delbruck. Briefly, a large number of cultures are allowed to grow in liquid medium starting from very low inocula (to avoid the presence of resistant mutants in the inoculum). The frequency of mutants present in each grown culture is determined and, using a statistical calculation, one can obtain the frequency of mutation.

4.2. Biochemical Bases of Resistance

An antibiotic inhibits bacterial growth if it (1) is able to reach the site of its action, (2) interacts with a structure involved in an essential function, and (3) substantially inhibits this function. A microorganism becomes resistant to an antibiotic if at least one of these steps is no longer operative.

This can result from one of the following main biochemical mechanisms:

1. Transformation of the antibiotic into an inactive form
2. Modification of the cell's target site for the antibiotic
3. Modification of the permeability of the microorganism to the antibiotic
4. Increased production of the structure inhibited by the antibiotic

Table 4.1 lists the types and the mechanisms of resistance for the more familiar antibiotics.

4.2.1. Transformation of the Antibiotic into an Inactive Form

The resistant strain produces an enzyme capable of chemically transforming the antibiotic into an inactive product. Among these enzymes, the most important are:

1. The *peptidases*, which hydrolyze peptide bonds, and in particular the *β-lactamases* that open the β-lactam ring of penicillins and cephalosporins. The β-lactamases produced by different resistant strains are not all identical. For example, the staphylococcal β-lactamase hydrolyzes penicillins but is not active against cephalosporins, contrary to that of *Escherichia coli* which is active on both classes.
2. The *acetyl-transferases*, which inactivate antibiotics by transferring an acetyl group from an acetyl donor to a functional group of the antibiotic. An example is the conversion of chloramphenicol to acetyl- or diacetyl-chloramphenicol.

Table 4.1. Mechanisms of Resistance to Some Representative Antibiotics

Antibiotic	Type of resistance	Location of genetic determinant	Description of resistance mechanism
β-Lactam antibiotics	1. Inactivation	Extrachromosomal and chromosomal	β-Lactamases that open the β-lactam ring. Some specific for penicillins or cephalosporins, others do not distinguish between the two antibiotics
	2. Alteration of site of action	Chromosomal	Resistance to all the β-lactams as a result of modified PBPs (*S. aureus*)
	3. Permeability	Variable	Alteration of porins
	4. Tolerance	Unknown	Inhibition of growth by low antibiotic concentrations but no bactericidal effect even at high concentrations
Chloramphenicol	1. Inactivation	Extrachromosomal	Acetylation by an inducible enzyme
	2. Modification of site of action	Chromosomal	Alteration of rRNA 23 S
Aminoglycosides	1. Inactivation	Extrachromosomal	*N*-Acetylation, phosphorylation, adenylation related to various inactivating enzymes
	2. Permeability	Chromosomal	Energy deficiency, modification of porins (?)
	3. Modification of site of action	Chromosomal	Modification of RNA (typical of microorganisms that produce the antibiotic)

Streptomycin	1. Modification of site of action	Chromosomal	Alteration of the S 12 protein in the ribosomal 30 S subunit and alteration of the 16 S rRNA
	2. Inactivation	Extrachromosomal	Analogous to that of the other aminoglycosides
Kasugamycin	Modification of site of action	Chromosomal	Alteration of the 16 S RNA of the ribosomal 30 S subunit
Erythromycin	1. Modification of site of action	Chromosomal and extrachromosomal	Alteration of proteins of the ribosomal 50 S subunit and methylation of the RNA
	2. Inactivation	Extrachromosomal	Hydrolysis of lactone and consequent opening of the ring
Rifamycins	Modification of site of action	Chromosomal	Alteration in the β subunit of RNA polymerase
Cycloserine	Alteration of permeability	Chromosomal	Modification of the transport system for D-alanine and glycine, used by cycloserine
Tetracyclines	Alteration of permeability	Chromosomal	Decreased transport efficiency
		Extrachromosomal	A specific protein increases the efflux from the cell
Fosfomycin	1. Alteration of permeability	Chromosomal	Modification in the transport system of glycerophosphate or glucose-6-phosphate (used to transport fosfomycin)
	2. Inactivation	Extrachromosomal	Inactivating intracellular enzyme (reaction not yet identified)

3. The *phosphoryl-transferases*, which covalently link phosphate group to the antibiotic. Streptomycin can undergo this modification.
4. The *adenyl-transferases*, which inactivate the antibiotic by transferring an adenyl group to it.

Inactivating enzymes can be produced constitutively or induced by the antibiotic. In the latter case, the cell that contains the gene for resistance possesses control mechanisms that repress its expression (e.g., the synthesis of an enzyme, such as β-lactamase) when there is no need for it, i.e., in the absence of the antibiotic. In the presence of the antibiotic, a mechanism called *induction* occurs and the cell synthesizes the inactivating enzyme. Note that the antibiotic does not cause resistance, but only induces the expression of resistance genes present in the cell. In this case, the genetically controlled inducibility is the difference between susceptible and resistant bacteria.

4.2.2. Modification of the Antibiotic Target Site

Many antibiotics act by inactivating a target protein that can be called generically a receptor. The antibiotic binds to the target forming a more or less stable complex. One large class of resistant mutants is comprised of bacteria that, through a mutation, develop a target protein unable to bind the antibiotic, or less often, a target that retains its function even after formation of the complex. Frequently, this difference consists of substitution of a single amino acid in the protein chain. Examples of this type are some mutants resistant to streptomycin with an alteration in the ribosomal protein S12, which contributes to the formation of the complex of streptomycin with the ribosomal RNA. Similarly, there are mutants resistant to erythromycin that have an alteration in protein L4 of the ribosomal subunit 50 S, which is the site of action of this antibiotic and of other macrolides. Rifampin-resistant bacteria are another example. DNA-dependent RNA polymerase (the target of rifamycins) isolated from these mutants is not capable of forming a complex with the antibiotic. In several cases it has been demonstrated that this is the result of an alteration of the β chain, one of the five proteins that make up the enzyme. Resistance to antibacterial quinolones can arise from specific mutation in the A or B subunits of the enzyme DNA gyrase.

Bacteria have also been isolated that are both resistant to and dependent on the presence of the antibiotic for growth, i.e., they do not grow if the antibiotic is not present. This curious finding has been explained by postulating that the resistance gene would give rise to a target protein, which is inactive in its native form but which becomes functional after forming a complex with the antibiotic, which acts as an allosteric effector. Examples of this situation are some laboratory strains of *Mycobacterium tuberculosis* that are dependent on streptomycin.

4.2.3. Change in Permeability of the Microorganism to the Antibiotic

Like other organic molecules, antibiotics penetrate the cell membrane by one of two major mechanisms, passive diffusion or specific active transport. When the physical properties of the antibiotic are compatible with passive diffusion, it is very difficult to produce a mutation that makes the cell impermeable to the antibiotic. This would in fact imply a major change in the membrane structure that most probably would be lethal to the cell. When penetration is the result of a specific transport mechanism, a specific carrier protein may be involved and an alteration of this permease may result in development of resistance. This is the case for fosfomycin, which is taken up into bacteria by the same permease that transports glycerophosphate, and for cycloserine, which is transported by the carrier of some d-amino acids.

In the case of gram-negative bacteria, as discussed in Chapter 3, the penetration of many substances through the outer membrane is related to the presence of particular proteins called porins. An alteration in expression or structure of these may decrease permeation and cause resistance, as has been demonstrated for various β-lactams, tetracyclines, and chloramphenicol.

4.2.4. Increased Production of the Enzyme Inhibited by the Antibiotic

Mutants of this type are quite frequent among organisms resistant to antimetabolites, such as 5-methyltryptophan or trimethoprim. In the first case more tryptophan is produced: in the second, more folic acid reductase. An analogous situation has been described for mutants resistant to cycloserine; they produce an increased amount of alanine racemase or D-alanyl-D-alanine synthetase, enzymes inhibited by the antibiotic.

4.2.5. General Considerations

From the preceding discussion, it is possible to draw certain general conclusions:

1. There are several biochemical mechanisms for resistance to antibiotics. Some of these (such as the second and fourth types de-

scribed above) are related to the mechanism of action of the antibiotic. Others are independent of it.

2. For any given antibiotic there may be strains that are resistant through different mechanisms. For example, resistance to streptomycin is in some strains related to an inactivating enzyme and in others to a modification of the site of action.
3. When two antibiotics act at the same site, cross-resistance is frequent. In fact, bacteria resistant because of modification of the site of action will usually be resistant to both antibiotics. However, when the resistance to one of the two antibiotics has a different basis (for instance, the presence of an inactivating enzyme) the other antibiotic may remain active.

4.3. Genetic Aspects

4.3.1. Role of the Antibiotic in Transformation of a Bacterial Population from Susceptible to Resistant

During the process of duplication of genetic material, errors of replication may occur that lead to a change in the genetic information of the microorganism. Such modifications are called *mutations*. Some mutations may render the microorganism resistant to an antibiotic to which the wild type was susceptible. Spontaneous mutations toward resistance usually occur with a low frequency, of the order of 10^{-7} to 10^{-10} or less, and therefore the fraction of resistant cells in a bacterial population is always very small. However, when the population is grown in the presence of the antibiotic, the susceptible bacteria are inhibited while the resistant ones continue to multiply, so that eventually the whole population is composed of resistant bacteria. This action of the antibiotic is termed *selection of resistant mutants*. The emergence and spread of bacterial populations resistant to antibiotics are the result of the combined effects of mutation and selection.

It can be shown experimentally that an antibiotic acts only as a "selecting agent" and not as a "mutagenic agent." This is done by showing that the incidence of resistant cells in a susceptible population is independent of the presence or absence of the antibiotic, which is to say that resistant mutants arise in a bacterial population before the population comes into contact with the antibiotic. Various investigators have used different experimental approaches to demonstrate this point. A simplified example of one of these approaches is presented and explained in Figure 4.1.

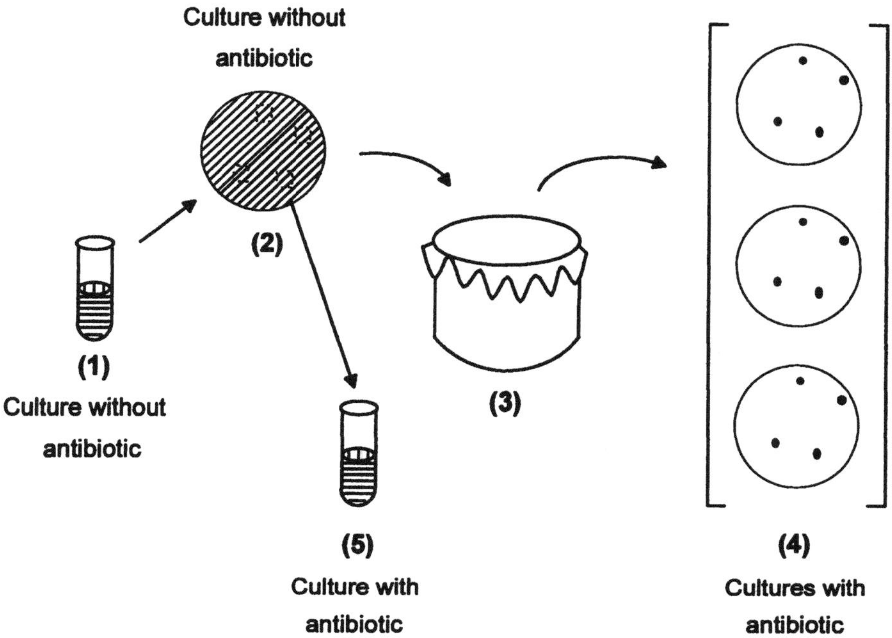

Figure 4.1. Bacteria from a culture of *E. coli* (1) susceptible to streptomycin are used to seed an agar plate (2) and are allowed to grow overnight. With the aid of a velvet plug (3), the plate is then replicated on a few agar plates (4) containing streptomycin. Here only bacteria resistant to the antibiotic can multiply and form colonies. The identical positions of these colonies on the replica plates demonstrate that the resistant bacteria were present in the original plate without antibiotic and were transferred during the replication process. In fact, resistant cells can be isolated from areas of the original plate corresponding to the position of the colonies, and identified by subculture in the presence of streptomycin (5).

4.3.2. Cell-to-Cell Transfer of Resistance

A resistant bacterium can in certain instances transfer the information for resistance to a susceptible one, causing it to become resistant and capable of further transferring the resistance to other bacteria. While it was initially believed that the process of mutation and selection was the main cause of resistance, it was later understood that the transfer of resistance is the main factor for the diffusion of resistance among the strains of clinical interest. The genetic information for resistance can be contained within either the DNA of the bacterial chromosome or small DNA molecules in the cytoplasm, called *plasmids*. The first case is called *chromosomal resistance*, the second *extrachromosomal resistance*.

4.3.3. Chromosomal Resistance

There are three mechanisms by which regions of the chromosome can be transferred from one bacterium to another:

1. *Conjugation*. This occurs when there is physical contact between two cells and a chromosome segment is replicated and transferred from one cell to the other. Once it has entered the recipient cell, the DNA from the donor can, by recombination, exchange with a homologous region of the recipient cell chromosome. When the segment transferred contains a *genetic determinant for resistance* to an antibiotic, both the donor and recipient bacteria will be resistant.
2. *Transduction*. This is the transfer, by a bacteriophage, of a segment of a bacterial chromosome from a donor strain to a recipient one. When a phage enters the lytic cycle, newly replicated phage chromosomes are packaged in the viral capsid, but a small proportion of particles include fragments of the host chromosome instead. These transducing particles can be absorbed by recipient cells where the DNA segment can recombine within the host chromosome. When the transduced segment contains a genetic determinant for resistance to an antibiotic, the result of the transduction will be transmission of resistance.
3. *Transformation*. This consists of the liberation into the environment of the DNA of a bacterium, as a consequence of cell lysis, and its uptake into another bacterium through specific receptors on the cell surface. In this case recombination can also occur. It has been demonstrated that this phenomenon, much exploited in the laboratory, also occurs in nature, but it appears of marginal importance in the transfer of resistance genes.

4.3.4. Extrachromosomal Resistance

The bacterial cytoplasm contains autonomous replication elements, called plasmids, that consist of circular molecules of double-stranded DNA. There are many types of plasmids. Some, called *episomes*, can be integrated in the chromosome, thus losing their replication autonomy. Other plasmids, termed *conjugative*, are able to move from one cell to another by a mechanism analogous to that of conjugation described above. They have generally a high molecular mass and are present in the cell in a low number of copies. Other plasmids, called *nonconjugative*, have a lower molecular mass and are present in a high number of copies.

Although unable to autonomously move from one cell to another, they can be transferred by the conjugation systems of other plasmids.

Plasmids are not essential for metabolism and growth of bacteria. However, they fulfill various special functions: some contain the genes for the synthesis of antibacterial toxins, such as the bacteriocins, others determine the production of enzymes able to degrade organic substances, and others confer virulence and pathogenicity to the host bacteria.

4.3.4.1. R Factors

Some plasmids contain genes determining resistance to antibiotics. When these plasmids also contain the genes for conjugation, and thus are transferable, they are called *R factors*.

Demonstration of the Existence of R Factors. Epidemiological and laboratory data gave rise to the suspicion that there may exist in bacteria extrachromosomal factors for multiple resistance capable of transferring resistance from strain to strain. In 1959, studies of an epidemic of intestinal shigellosis (bacillary dysentery) in Japan confirmed this suspicion. After antibiotic treatment, patients who had first excreted susceptible bacteria in their stools began to excrete bacteria resistant to four different antibiotics (multiple resistance). It would be very difficult to explain the appearance of multiple resistance by mutation. Spontaneous mutation toward resistance to a single antibiotic generally occurs with a frequency lower than 10^{-7}. The frequency of appearance of four independent mutations conferring resistance to four different antibiotics would be the product of the frequency of appearance of each mutation, i.e., 10^{-28} or lower, a very rare event indeed. The multiresistant bacteria isolated during the Japanese epidemic were too numerous and appeared too quickly after antibiotic treatment to be explained by independent mutations. In addition, how could the administered antibiotic have selected multiresistant bacteria and not mainly those resistant to itself?

The multiresistant bacterial strains isolated from patients treated with the antibiotic were mixed in test tubes with susceptible bacteria, and it was found that these acquired multiple resistance. It was known at that time that the individual chromosomal genes responsible for resistance to the four antibiotics were located in different positions along the chromosome. Therefore, simultaneous transfer would have required transfer of almost the entire chromosome. This never occurs in nature, since relatively small segments of chromosome are transferred in conjugation, transduction, and transformation. Therefore, it was postulated that there must exist genes for resistance to the four antibiotics that were not the same as those present in the chromosome and that were concentrated in small extrachromosomal elements (plasmids) capable of transfer from a bacterium to another. These plasmids were named R factors.

Physical and Genetic Nature of R Factors. By the classical techniques of molecular biology, it was shown that the R factors are double-helical DNA in the form of a covalently closed ring, with molecular weights of about 10^6–10^8. Their replication is coordinated with that of the chromosome, as shown by the fact that they are present in the cell in a 1:1 ratio with the chromosome. They are made up of genetic sequences responsible for their replication and transmission, and some resistance determinant genes.

Distribution of R Factors. R factors are found in many species of gram-negative bacteria, especially those of the Enterobacteriaceae (bacilli of the mammalian gastrointestinal tract). They can be transferred between bacteria of different species, e.g., between *Escherichia coli* and *Shigella*. They are present in both pathogenic and nonpathogenic organisms. They can therefore be found in the normal intestinal flora, which can serve as a reservoir for resistance, and transfer it to pathogenic bacteria.

4.3.5. The Role of Transposons

The classic mechanisms of chromosomal and plasmid DNA transfer and recombination are not sufficient to explain the mobility of the genes conferring resistance and the frequency at which chromosomal genes are inserted into plasmids. In fact, recombination usually takes place only between DNA regions that possess a high degree of homology. Acquisition of genetic material by plasmids and chromosomes can occur, independently from classical recombination, through transposable genetic elements, termed *transposons*, which are components of the genome of many organisms.

Transposons are DNA segments characterized at their two ends by short terminal sequences, called *repeats*, usually having an inverted orientation. The main property of transposons is their ability to easily insert themselves at different locations in either chromosomal or extrachromosomal DNA. They may contain only the genetic information necessary for their transposition, and in this case they are called *insertion*

Table 4.2. Properties of Some Resistance Transposons

Transposon	Size (kilobases)	Coded resistance
Gram-negative bacteria		
Tn1	5.0	Ampicillin
Tn2	5.0	Ampicillin
Tn3	5.0	Ampicillin
Tn5	5.1	Kanamycin
Tn7	14.0	Streptomycin, trimethoprim
Tn9	2.7	Chloramphenicol
Tn10	9.3	Tetracycline
Tn21	19.0	Streptomycin, sulfonamides
Tn1721	11.2	Tetracycline
Staphylococci		
Tn551	5.3	Macrolides, lincosamides, streptogramins
Tn552	6.1	Penicillin
Tn554	6.7	Macrolides, lincosamides, streptogramins, spectinomycin
Tn4001	4.7	Gentamicin, trimethoprim, kanamycin
Tn4002	6.7	Penicillin
Tn4003	3.6	Trimethoprim
Tn4201	6.6	Penicillin
Tn4291	7.5	Methicillin
Streptococci		
Tn916	16.4	Tetracycline

sequences, or they may contain additional genetic information, which, in several cases, consists of resistance determinants (see Table 4.2).

Because of this ability for transposition and insertion, transposons are the true factors responsible for the reorganization of the genetic information and, consequently, for the clustering of several resistance genes in a single plasmid.

4.3.6. Plasmids in Gram-Positive Bacteria

R factors, as already mentioned, are typical plasmids of gram-negative bacteria. However, the transfer of the resistance genes by conjugation is also operative in gram-positive bacteria. In streptococci, two groups of plasmids have been observed that are able to transfer genes. Conjugation in the absence of plasmids has also been observed. This latter process is believed to be determined by special conjugative transposons. The transfer of resistance genes by phage transduction in staphylococci is certainly important, but in this genus a form of conjugation is also present that has not yet been clarified. In enterococci, it was observed that genes are transferred with high frequency by plasmids or transposons through conjugation mediated by polypeptides called *sexual pheromones*. Furthermore, although at low frequency, conjugation between enterococci and staphylococci or streptococci has been observed.

4.4. Interesting Examples of Bacterial Resistance

4.4.1. β-Lactam Antibiotics

Resistance to the β-lactam antibiotics can be caused by three different mechanisms: (1) synthesis of inactivating enzymes, in particular, the β-lactamases, (2) reduction of permeability, and (3) modification of one or more proteins with which they bind (PBP). There is also a particular type of resistance, called *tolerance*: the antibiotic loses its bactericidal activity but maintains the bacteriostatic one.

4.4.1.1. Production of β-Lactamases

The β-lactamases are enzymes that hydrolyze the cyclic amide of the β-lactam molecule. In gram-negative bacteria, these enzymes are localized in the periplasmic region and inactivate the antibiotic as soon as it penetrates through the outer membrane. In gram-positive bacteria, they are excreted mainly into the culture medium. Consequently, while in gram-negatives resistance refers to single cells, in gram-positives this is rather a population phenomenon, as the enzyme produced by the resistant cells can protect, by inactivating the antibiotic in the culture medium, the susceptible bacteria also. The bacterial β-lactamases are extremely heterogeneous, both in structure and in substrate specificity. Most contain serine as the essential amino acid at the active site but some are known in which a different amino acid is involved in catalytic activity. Some β-lactamases are metalloenzymes.

β-Lactamases can be classified according to: substrate specificity (penicillinases, cephalosporinases, etc.), susceptibility toward inhibitors (isoxazolyl-penicillins, clavulanic acid), origin (chromosomal or plasmidic), and producing microorganisms. The principal characteristics of β-lactamases are presented in Table 4.3.

A convenient classification of β-lactamases is based on their origin, either chromosomal or plasmidic. Chromosomal β-lactamases are present in practically all gram-negative species. However, in some species

Table 4.3. Classification of β-Lactamases[a]

			Inhibitors			
Group	Class	Preferred substrate	Clavulanic acid	Isoxazolyl penicillins	Origin	Typical producers
1	I	Cephalosporins	No	Yes	Chromosomal	*Enterobacter, Proteus, Pseudomonas*
2a	II	Penicillins	Yes	No	Plasmids	*S. aureus*, rarely *E. coli*
2b	III and IV	Broad spectrum	Yes	Yes	Plasmids (TEM and SHV)	*E. coli, Haemophilus, Neisseria, Salmonella, Shigella, Pseudomonas*
2c	V	Penicillins (carbenicillin)	Yes	No	Plasmids (PSE)	*Pseudomonas, Proteus mirabilis*
2d	V	Penicillins (cloxacillin)	Yes	No	Plasmids (OXA)	*E. coli, Pseudomonas, Salmonella*
2e	—	Cephalosporins	Yes	Yes	Not determined	*Bacteroides, Proteus*
3	—	Various	No	No	Chromosomal	Metalloenzymes from *B. cereus* and *Flavobacter*
4	—	Penicillins	No	Var.	Chromosomal	Variable

[a] The "group" refers to the classification of Bush, the "class" to the classification of Richmond and Syke.

such as *E. coli*, their level is always low, so that, for instance, ampicillin that is sensitive to their action is, in practice, not inactivated. In contrast, in other pathogens such as *Pseudomonas*, *Enterobacter*, *Serratia*, and *Citrobacter*, these enzymess are inducible, i.e., in the presence of some β-lactams the expression of the enzyme rapidly increases and the antibiotic is inactivated. Most cephalosporins and ureido-penicillins are not inducers and are therefore unaffected by this type of resistance. However, as observed with cefamandole or azthreonam, cephalosporins and less frequently ureido-penicillins or monobactams can select derepressed mutants, which constitutively produce high levels of β-lactamase even in the absence of an inducer, and are thus resistant to all β-lactams.

Plasmid-mediated β-lactamases of gram-positive bacteria are almost exclusively produced by staphylococci, although, being transferable, they may be found in other species. Four types of these enzymes can be distinguished, of which types A, B, and C are inducible, whereas type D is constitutive. All hydrolyze penicillins preferentially but are ineffective against methicillin and isoxazolyl-penicillins.

In gram-negatives, plasmid-determined β-lactamases of the TEM type are widespread. TEM-1 and TEM-2 enzymes (the two proteins differ in only one amino acid) are almost equally active against benzylpenicillin and cephaloridine, but inactive against more recent cephalosporins. TEM-3 β-lactamases as well as some evolved from other plasmids (SHV-1 and PSE-29) are considered to be extended-spectrum β-lactamases, in that they inactivate most cephalosporins. However, these enzymes are not effective against cephamycins and moxalactam nor against combinations of a β-lactam antibiotic and a β-lactamase inhibitor. Several other plasmid-determined β-lactamases have been detected in recent years that we may call exotic β-lactamases. These are less clinically relevant, since they are produced by a few strains with a limited geographical distribution.

4.4.1.2. Reduction of Permeability

To exert their action, β-lactam antibiotics must reach the surface of the cytoplasmic membrane. To do so they must cross structures that are different in gram-positive and gram-negative bacteria.

In gram-positives the only barrier is the cell wall, which is usually permeable to small molecules such as β-lactams. In contrast, in gram-negatives they have to cross the outer membrane, through the porin channels, and the periplasmic space, where they may be inactivated by β-lactamases. The number of antibiotic molecules that reach the target on the cytoplasmic membrane depends on the rate of permeation through the pores of the outer membrane and the rate of inactivation in the periplasmic space. A mutation that results in reduction of permeation rate through the outer membrane will also result, at constant inactivation rate, in a smaller number of molecules reaching the target, thus effectively giving rise to resistance.

4.4.1.3. Modification of the Penicillin Binding Proteins

Resistance can be caused by insensitivity of the target, i.e., the production of modified proteins involved in the peptidoglycan synthesis unable to covalently bind β-lactam antibiotics. Two clinically very important variants of this type of resistance are recognized.

Mutant strains of the genus *Staphylococcus* produce a PBP (PBP-2′) remarkably different from those produced by susceptible strains, which can functionally substitute for PBP-2 and PBP-3 and has a very low affinity for any of the known β-lactam antibiotics. These strains are called methicillin-resistant *Staphylococcus aureus* (MRSA), as methicillin is the prototype of the antibiotics active against the strains resistant by production of penicillinases. These mutants are resistant to all known β-lactam antibiotics, and most often carry determinants for resistance to aminoglycosides, macrolides, and tetracyclines. This type of resistance is determined by a transposon designated Tn4291. A similar situation is found in resistant *Enterococcus* strains, which produce larger amounts of another protein, PBP-5, also insensitive to β-lactams.

In resistant mutants of some species, mosaic PBP encoding genes have been detected, in which portions of PBP genes of an insensitive species are inserted into the gene coding for a normal PBP. The protein expressed are therefore altered PBPs with low affinity for β-lactams. As might be expected, this type of resistance is found only in species having a natural competence for transformation, such as pathogenic *Neisseria* strains that acquired resistance from *Nocardia lactamica*, or resistant *Streptococcus pneumoniae* strains that possess genes derived from *Streptococcus mitis*.

4.4.1.4. Tolerance toward β-Lactam Antibiotics

A particular type of resistance to β-lactam antibiotics, called *tolerance*, has been observed in recent years in clinical isolates. Tolerant bacteria are inhibited but not killed by normal antibiotic concentrations or even by higher concentrations.

The biochemical basis of tolerance does not appear to be an alteration of PBPs, or the production of β-lactamases, but rather an alteration of the cell autolysins, which would lose their ability to induce the lethality mechanism. Tolerance has been observed mainly in clinical isolates of pneumococci and of other gram-positive bacteria.

4.4.2. Macrolides

4.4.2.1. Common Mechanism of Resistance

Erythromycin acts by binding to a specific region of the ribosomal 23 S RNA (see Chapter 3). This site is shared, at least in part, by other macrolide antibiotics and also lincomycin and streptogramin. Normally, ribosomes from *Staphylococcus* are highly sensitive to erythromycin but resistant strains have been increasingly found whose ribosomal affinity for the antibiotic was substantially reduced. This suggested that resistance involves some form of ribosomal modification. Comparison of ribosomal RNAs of sensitive and resistant strains revealed that 23 S rRNA of the latter was methylated, and more exactly that the difference consisted of *N,N*-dimethylation of the adenine base at nucleotide position 2058 (according to the *E. coli* numbering system).

The genes determining this type of resistance are present in several bacterial species but of practical interest are only those originating from *S. aureus*, namely *ermA* on transposon Tn554 and *ermC* on plasmid pE194. Both these genes code for similar adenosylmethionine-dependent methylases. The activity of these enzymes is inducible by erythromycin, oleandomycin, celesticin (a lincomycin derivative) but normally not by 16-membered macrolides or lincomycin. However, the induced strains become resistant to all *m*acrolides, *l*incosamides, and *s*treptogramins, and therefore this type of resistance is called MLS resistance. Constitutive rather than inducible mutants can be selected by plating *S. aureus* cells at high density on a 16-membered macrolide such as spectinomycin.

An interesting aspect of the mechanism of this induction is that it does not consist of the classical derepression of transcription but of posttranslational modification of the protein.

4.4.2.2. Specific Mechanisms of Resistance to Erythromycin

In addition to the common mechanism described above, resistance to erythromycin can result from an alteration of the chromosomal gene for ribosomal protein L4, whose modification prevents the binding of the antibiotic to the ribosome. Other chromosomal mutations are known that may confer resistance to erythromycin but their connections with the ribosome components have not yet been clarified.

Another resistance mechanism, mediated by plasmids, involves the production of inactivating enzymes. Two esterases have been identified that open the macrolide ring by hydrolyzing the lactone bond.

4.4.3. Aminoglycosides (Aminocyclitols)

4.4.3.1. Streptomycin

As discussed in Chapter 3, the mechanism of action of streptomycin differs from that of other aminoglycosides and thus the mechanism of resistance also differs. A specific mechanism, determined by a chromosomal mutation involves substitution of one amino acid in the S12 protein of the ribosomal 30 S subunit, which becomes unable to bind the antibiotic. It is noteworthy that alterations of ribosomal proteins S4 or S5 result in hypersensitivity to the antibiotic and some specific mutations in ribosomal proteins S12 or S8 can lead to growth dependence on aminoglycosides.

A second type of resistance involves the production of inactivating enzymes whose genes are borne by plasmids. In particular, adenylation of the hydroxyl in position 6 of streptomycin occurs in staphylococci, and of the hydroxyl in position 3″ in gram-negatives. Resistant strains of both gram-positives and gram-negatives exist that phosphorylate the hydroxyl in position 3′. The ability to phosphorylate the hydroxyl in position 6 is typical of *Pseudomonas* strains.

4.4.3.2. Aminoglycosides Containing 2-Deoxystreptamine

A number of experimental observations indicate that the 3′ terminal region of 16 S rRNA is involved in binding of aminoacyl tRNAs, and that aminoglycoside antibiotics interfere with this binding.

Many attempts have been made to isolate mutants showing one-step resistance to these antibiotics, with the aim of identifying the precise antibiotic binding site on the ribosome. The fact that these attempts have been unsuccessful may relate to two possible reasons: (1) there are multiple binding sites for these antibiotics, (2) the antibiotic does not interact with the proteins but directly with ribosomal RNA. As each ribosomal RNA is coded by several identical genes, the mutation of a single gene would be ineffective (recessive mutation). In this respect, it is noteworthy that the microorganisms producing these antibiotics protect themselves from the action of the antibiotic through methylation of a specific base of the 16 S rRNA.

The main mechanism of resistance to these aminoglycosides involves inactivation by enzymes of plasmid origin. Gentamicin is inactivated through phosphorylation of the hydroxyl at position 2″ in gram-positive resistant strains, and through adenylation of the same hydroxyl in gram-negatives. In the latter species, acetylation of the amine at position 3 also has been frequently observed.

4.4.3.3. Decreased Uptake

Aminoglycoside antibiotics enter the cell through a complex transport process whose mechanism is still a matter of controversy. There is, however, agreement on the fact that the main driving force for uptake through the cytoplasmic membrane is the internal negative electric potential. This explains some forms of phenotypic resistance, such as inactivity at low pHs, conditions under which basic antibiotics are protonated. It can also explain the insusceptibility of anaerobic organisms (or of facultative aerobes grown under anaerobic conditions) which would not possess enough negative potential for uptake.

Similarly, clinical isolates have been reported whose resistance to aminoglycosides is associated with a variety of "energy" deficiencies. It has been suggested that their clinical relevance may be underrated, because they are generally slow-growing and therefore in susceptibility laboratory tests their presence can be hidden by fast-growing sensitive revertants.

Chapter 5

Activity of Antibiotics in Relation to Their Structure

Since it was observed that different natural penicillins display different biological properties, the study of the relationship between *chemical structure* and *microbiological activity* in an antibiotic molecule has become of major interest.

This study involves a systematic modification of the antibiotic molecule to establish which *structural elements* or *chemical functions* are necessary for microbiological activity, and which are dispensable.

Many structure–activity relationships have been established by empirical modifications of the natural molecules with the practical aim of introducing desirable characteristics (e.g., higher activity, activity on resistant strains, acid stability) or removing undesirable ones (e.g., toxicity, excessive protein binding).

As data on the biological effects of these chemical modifications became available, structure–activity patterns began to emerge, which allowed a more direct and rational approach to chemical transformation. Accordingly, in describing the structure–activity relationships for the different families of antibiotics, we present first the initial empirical phase and follow with a concise description of the current knowledge on these relationships.

In the section "Unresolved Problems" some biological aspects (e.g., spread of resistance, side effects) are discussed whose solutions are the current target of semisynthetic derivative design.

The traditional chemical modification approach, both in its empirical and rational form, has been very successful as evidenced by the numerous semisynthetic antibiotics introduced into clinical practice. More recently, with the aid of powerful computers, it has become possible to directly visualize the interaction between the antibiotic molecule and the macromolecule that constitutes its *biological target*. This requires a precise knowledge of the target molecule, which at present is unlikely.

5.1. Introduction

5.1.1. Basis of the Structure-Activity Relationship

At the molecular level, the activity of an antibiotic can be described by its ability to form a complex with a molecule essential for the growth of the bacterial cell, thus inhibiting its function. Antibiotics can bind with molecules of various natures. Often they are proteins endowed with enzymatic activity; the examples of interaction with DNA are quite common, while interactions with ribosomal RNA were revealed more recently. In some rare cases complexation occurs with small molecules that are substrates of the enzymatic activity. Some antibiotics interfere with complex biological structures such as the cell membrane.

In some cases, a covalent bond is formed between the antibiotic and the target molecule, but in the majority of the cases, complexes are formed through so-called *weak bonds* between specific chemical groups of the antibiotic and the macromolecule. The energy of any single one of such bonds is normally not sufficient to ensure the formation of a stable complex. Therefore, if inhibition has to occur, several weak bonds must form between functional groups of the antibiotic and those of the macromolecule. As the strength of a weak bond is dependent on the distance between the interacting atoms, several bonds of sufficient strength can form between the two molecules only when the spatial structure of the antibiotic allows several functional groups to come within the appropriate distance.

Therefore, we can look at the antibiotic molecule as a rigid structure, a framework that carries and maintains some functional groups able to interact with the macromolecule in their correct positions. Also in cases in which a covalent bond is formed, the general antibiotic structure and the interaction of functional groups determine the correct positioning of the reactive site, and thus the formation of the covalent bond. It is intuitive that in a very small molecule any change in the structure will cause a change in the position of the groups participating in the formation of the bonds, with a consequent loss of activity. Conversely, in larger molecules changes can be produced that do not affect these functional groups or their relative positions.

In conclusion, the study of the relationships between chemical structure and biological activity of an antibiotic consists in identifying the functional groups directly involved in activity at the molecular level and determining which changes can be made in the molecule without causing loss of activity.

5.1.2. Chemical Modifications of Natural Antibiotics

Changes in the molecule that do *not* interfere with the activity at the molecular level, i.e., with the ability to form a complex with a macromolecule in the bacterial cell, can be of fundamental importance with regard to *in vivo* biological activity and practical usefulness of the antibiotic. Properties such as transport across the membrane, and hence penetration into the bacterial cell, absorption and distribution throughout the body, and susceptibility to the action of inactivating enzymes can be profoundly influenced by more or less drastic modifications in the original molecule.

During the 1950s, there was considerable skepticism about the possibility of substantially improving antibiotic efficacy by chemical modification of natural products. This was in part based on the lack of success of early attempts made on the first antibiotics such as streptomycin and actinomycin. In addition, there was a widespread belief that it would be impossible to improve the activity of a natural product.

The first argument was obviously not valid, as it was based on a few unsuccessful examples. As for the belief that nature could not be improved, it is admittedly difficult to improve a product that is the result of selective evolution over millions of years. In fact, very few examples are known in which the affinity of an antibiotic for its target molecule has been increased by chemical modification. However, in the case of an antibiotic, the evolutionary selection was based in one way or another on its contribution to the survival of the producing microorganism in its natural habitat, and certainly not on its therapeutic activity.

The results discussed in the following sections amply demonstrate the advantages that can be obtained by chemical modification of antibiotics.

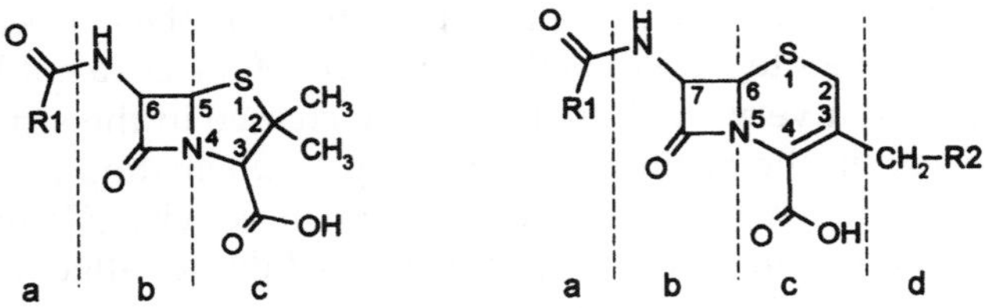

Penicillins

a) side chain

b) ß-lactam ring

c) thiazolidine ring

Cephalosporins

a) side chain

b) ß-lactam ring

c) dihydrothiazine ring

d) group at position 3

Figure 5.1. β-Lactam antibiotics.

5.2. β-Lactam Antibiotics*

The antibiotics of this class are characterized chemically by the presence in the molecule of a β-lactam, a cyclic amide forming a four-atom ring. The integrity of this ring is essential for activity; in fact, the opening of this ring, which can occur rather easily either chemically or enzymatically, is both the main difficulty encountered in the chemical manipulation of the molecule and the most common cause of biological inactivation by bacterial enzymes. All β-lactam antibiotics have a common mechanism of action: inhibition of the bacterial cell wall synthesis by interacting, in different ways, with the penicillin binding proteins (PBP), enzymes involved in the synthesis of peptidoglycan (see Section 3.3).

The β-lactam group includes two of the most important families of antibiotics, the *penicillins*, in which the β-lactam ring is fused with a thiazolidine (a five-atom ring), and the *cephalosporins*, in which there is a dihydrothiazine (a six-atom ring) (Figure 5.1). Most of the derivatives currently used in therapy have been obtained by modification of the

*The minimal inhibitory concentrations reported in this and the following sections have been compiled from different literature sources and therefore are not homogeneous. Normally, they refer to susceptible laboratory strains because, except for particular cases, the activity against these is the basis for the studies on structure–activity relationship. These MICs should not be considered as an indication for therapy since, in this respect, the MIC_{50} and MIC_{90}, which vary with time and environment, have to be considered.

lateral chain at position 6 of penicillins or position 7 of cephalosporins. Active derivatives of the latter have also been obtained by modification of the substituent at position 3.

In addition to these "classic" β-lactam antibiotics, several new and interesting β-lactam structures have been isolated from microorganisms or obtained by semisynthesis. Collectively referred to as "nonclassic" β-lactams, they include the *cephamycins* or 7-methoxycephalosporins, the *carbapenems*, and the *oxapenems*, in which the five-atom ring contains a carbon or an oxygen atom instead of sulfur, and *monobactams*, in which the β-lactam ring is isolated.

5.2.1. Natural Penicillins

The era of extensive use of antibiotics in medicine began in 1942 when penicillin G (Figure 5.2) was introduced into clinical practice. Penicillin G is active against gram-positive bacteria, *Neisseria* and *Treponema pallidum*, the causative agent of syphilis, while it is scarcely active against gram-negative bacilli. Its great efficacy *in vivo* and lack of toxicity even at very high doses have made it for half a century the antibiotic of choice in the treatment of several infectious diseases. The pharmacokinetic properties of penicillin G, however, are not satisfactory. It is absorbed only partially when administered orally and most of it is inactivated by the acidic pH in the stomach. When injected, e.g., as the soluble potassium salt, it is rapidly adsorbed but also rapidly excreted in the urine, with a serum half-life of only 30 min. However, the insoluble salts, such as procaine penicillin or benzathine penicillin, are excreted very slowly.

Penicillin G was chosen among the natural penicillins produced by *Penicillium notatum* (Figure 5.2) because its production could be substantially increased by the addition of the lateral chain precursor phenylacetic acid, and because it was more effective in curing bacterial infections in animals, although penicillin K appeared to be more active *in vitro*. This revealed an important aspect of structure–activity relationship in penicillins: when the lipophilicity of the side chain is increased, the binding to the serum proteins is also increased, normally resulting in lower therapeutic efficacy.

The following objectives of the modification of penicillin G were soon identified:

1. Improvement of absorption after oral administration
2. Enlargement of the spectrum of activity to gram-negative bacteria
3. Reduction of the incidence of allergic reactions
4. Acquisition of activity against resistant staphylococcal strains

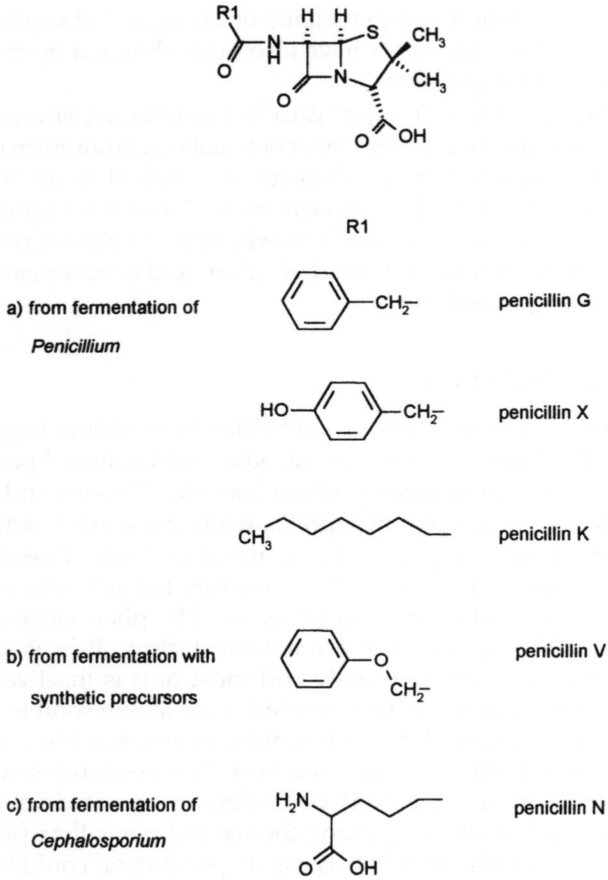

Figure 5.2. Penicillins obtained by fermentation.

The last two objectives arose from the clinical observations that allergy to penicillin G was common and sometimes severe, and that, with the widespread use of the drug, resistant strains were rapidly emerging.

5.2.2. Penicillins Obtained by Fermentation with Synthetic Precursors

The existence of several natural active penicillins differing only in the structures of their side chains suggested that other derivatives with improved characteristics could be obtained by introducing further variations into the chain. However, for several years no chemical method was available to produce these variations.

After the observation that the production of a given penicillin could be increased by addition to the cultures of the acid corresponding to its side chain (e.g., phenylacetic acid increases the production of penicillin G), attempts were made to biosynthesize new penicillins by adding nonnatural precursors to the fermentation. Some of these were utilized by the microorganism and gave rise to new penicillins, including *penicillin V* (Figure 5.2) or *phenoxymethylpenicillin,* obtained after addition of phenoxyacetic acid. It is orally active since it is more resistant than penicillin G to acid degradation.

However, this method showed a limited potential because of the structural requirement that compounds be accepted as substrates by the enzymatic system of penicillin biosynthesis. Practically, only substituted phenyl or phenoxyacetic acids and some α,ω-dicarboxylic acids were utilized. None of the products obtained, with the exception of the abovementioned penicillin V, offered any advantages over penicillin G.

5.2.3. 6-Aminopenicillanic Acid

The breakthrough that led to the preparation of thousands of semisynthetic penicillins was the isolation of 6-aminopenicillanic acid. This product, which from a chemical standpoint is the penicillin nucleus, was first isolated from fermentation of *Penicillium* to which no side chain precursors had been added. Later, it was found that several microbial enzymes could split benzylpenicillin into phenylacetic acid and 6-aminopenicillanic acid. This reaction is still used industrially, although later chemical methods of hydrolysis have been developed. Semisynthetic penicillins can now be prepared by addition of the appropriate acid chloride or acid anhydride to 6-aminopenicillanic acid, according to the scheme shown in Figure 5.3. Alternative methods of acylation are also available, such as the use of azides and the condensation of the acids with dicyclohexylcarbodiimide. Enzymatic methods are also known.

5.2.4. Penicillins Absorbed after Oral Administration

Starting from 6-aminopenicillanic acid, many structural analogues of penicillin V have been synthesized. Among these, *phenethicillin* was the first semisynthetic penicillin used in therapy, followed by its homologue *propicillin* (Figure 5.4). The spectrum of antimicrobial activity of these compounds is generally similar to that of penicillin G. They are,

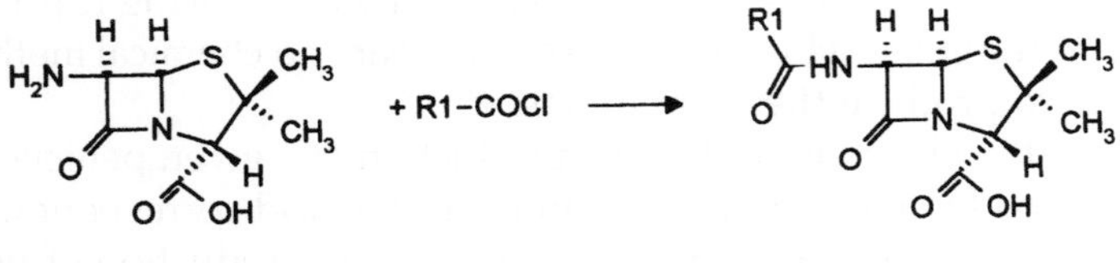

6-Aminopenicillanic acid

Figure 5.3. Scheme of synthesis of semisynthetic penicillins.

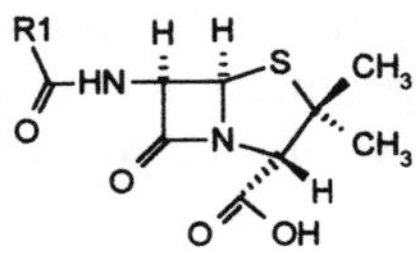

R1	Name	MIC (µg/mL) *S.aureus*	*S.pyogenes*	*E.coli*	*N.gonorrhoeae*	*H.influenzae*	Oral absorption *
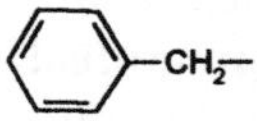	Penicillin G	0.03	0.008	64	0.008	1	100
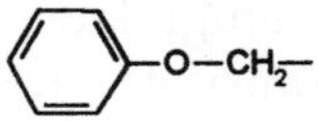	Penicillin V	0.03	0.02	128	0.003	4	250
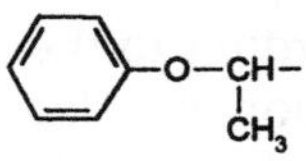	Phenethicillin	0.03	0.03	>200	0.1	4	400
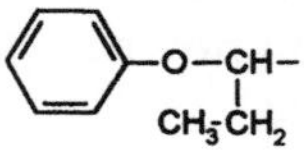	Propicillin	0.06	0.03	>200	-	-	400
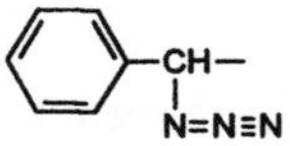	Azidocillin	0.04	0.01	-	-	0.8	300

* peak serum levels in man, percent of those observed with penicillin G

Figure 5.4. Acid-resistant penicillins. *In vitro* activity and oral absorption in comparison with penicillin G.

however, less active against *Neisseria* and thus are not recommended for the treatment of gonorrhea.

Their advantage over penicillin G relates to better absorption when given orally, resulting from: (1) stability in acid medium and (2) increased lipophilicity of the side chain. Both experimental evidence and theoretical considerations indicate that the stability of the penicillin nucleus toward acids is influenced by the chemical nature of the chain: the presence of an electron-attracting moiety in the position α to the amide increases the stability. In penicillin V and its homologues, the electron-attracting moiety is the oxyphenyl group. Increased lipophilicity of the chain directly influences absorption in the intestine, as is the case for many drugs. And, as previously mentioned, the extent of binding with serum proteins is also increased, so that the presence of long lipophilic chains may result in decreased efficacy.

Azidocillin, a new penicillin well absorbed orally, was subsequently proposed for clinical use. Although its spectrum of activity is generally similar to that of phenoxypenicillins, azidocillin appears to be more active against *Haemophilus influenzae.*

Other orally absorbed penicillins that possess specific properties such as activity against resistant *S. aureus* strains or an extended spectrum of action are discussed in the following sections.

5.2.5. Penicillins Insensitive to Staphylococcal Penicillinase

With the extensive use of penicillin G, *Staphylococcus* strains resistant to the antibiotic emerged. Apparently this was the result of these strains' ability to produce enzymes called β-lactamases, or penicillinases, that inactivate penicillin by the reaction shown in Figure 5.5 (see also Section 4.4.1.1).

The synthesis of a number of penicillins with structural variations in the side chain revealed that when the carbon atom α to the amide is included in an aromatic ring carrying substituents in the *ortho* position, the resulting steric hindrance protects the nearby lactam ring from the enzymatic attack. The first derivative used clinically for the treatment of infections caused by penicillinase-producing staphylococci was *methicillin*. Since it has low antibacterial activity and is inactivated by acids, it must be administered by injection in large doses (several grams per day) (Figure 5.6).

Similar properties but higher antimicrobial activity are present in *nafcillin,* another penicillinase-resistant penicillin mainly used in the United States. Activity against penicillinase-producing staphylococci and insensitivity to acids (and thus absorption after oral administration)

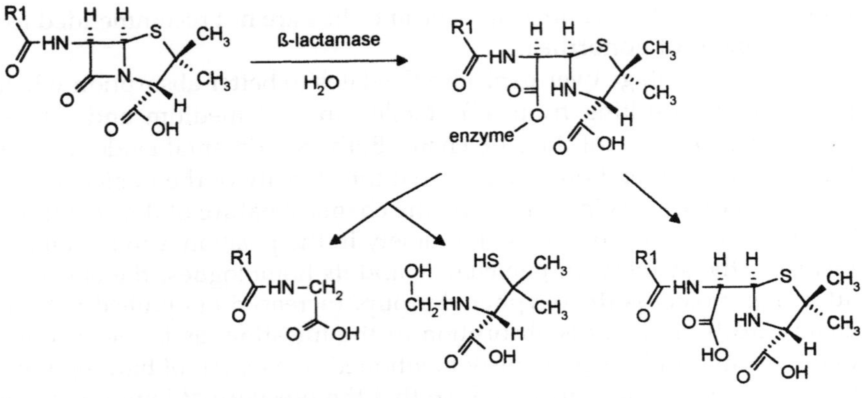

Figure 5.5. Enzymatic degradation of penicillins.

are combined in isoxazolylpenicillins. They are *oxacillin*, whose serum half-life is very short, and its halogen derivatives *cloxacillin*, *dicloxacillin*, and *flucloxacillin*, which are better absorbed and more slowly eliminated. The longer-lasting serum levels are partially the result of a greater degree of serum protein binding. As earlier discussed, this may result in reduced efficacy. Although a quantitative assessment of the significance of these different parameters is difficult, it may be noted that the isoxazolylpenicillins are about ten times more active *in vitro* than methicillin, but their therapeutic dosages *in vivo* are similar to that of methicillin.

With extensive clinical use, strains of *Staphylococcus aureus* resistant to both methicillin and isoxazolylpenicillins have emerged. These strains are not penicillinase producers, but have a modified PBP that is insensitive to the action of all known penicillins (methicillin resistance, see Section 4.4.1.3).

5.2.6. Penicillins with Extended Spectra of Activity

A large research effort has been devoted to the synthesis of penicillins potentially active against gram-negative strains that are practically insusceptible to penicillin G and to the other derivatives so far described. This research was initially directed by the observation that both penicillin N (see Figure 5.2), a product of all cephalosporin-producing organisms, and *p*-aminobenzylpenicillin, a semisynthetic derivative, had some activity against gram-negative bacteria. By comparing the activities of the many derivatives prepared, it has been possible to establish a correlation between some structural features and the property of inhibiting the

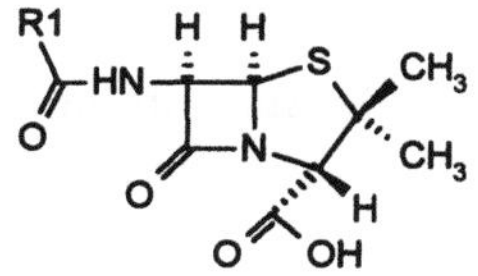

R1	Name	MIC (µg/mL)			Oral absorption *
		S.aureus	*S.aureus* (penicillinase producer)	*S.pyogenes*	
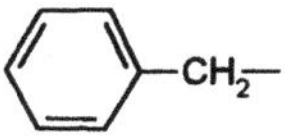	Penicillin G	0.03	125	0.008	100
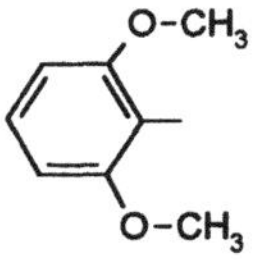	Methicillin	1	2	0.2	100
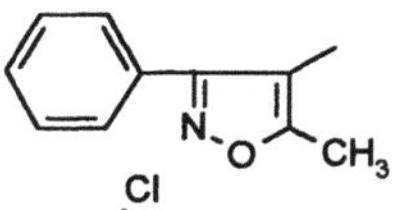	Oxacillin	0.4	0.4	0.1	200
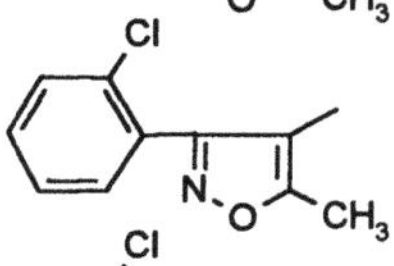	Cloxacillin	0.2	0.3	0.1	350
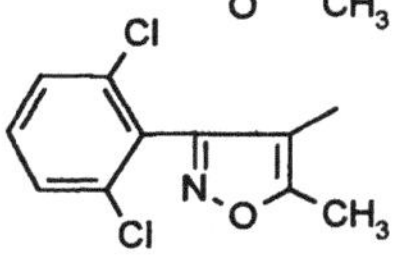	Dicloxacillin	0.06	0.1	0.05	700
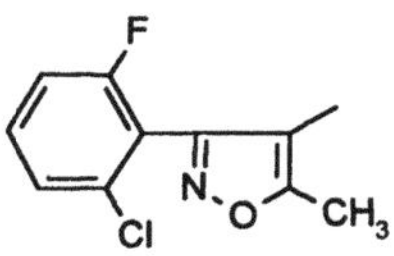	Flucloxacillin	0.1	0.3	0.05	750
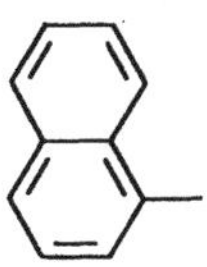	Nafcillin	0.3	0.3	0.03	100

* peak serum levels in man, percent of those observed with penicillin G

Figure 5.6. Penicillins active against penicillinase-producing staphylococci.

growth of *E. coli* and other gram-negative strains. This correlation can be summarized as follows:

1. A moderate enhancement of activity is obtained by substitution of the phenyl group of penicillin G with certain heterocyclic rings.
2. The activity decreases when the lipophilic character of the chain is increased.
3. The effect of polar substituents on the chain in positions far from the amide bond is positive but small. When the substituent is an amino group this effect is more pronounced.
4. The presence of a polar group α to the amide considerably increases the activity. In this case, the stereochemical configuration is also important.
5. The activity decreases when the carbon atom α to the amide is fully substituted.

Among the many derivatives prepared, *ampicillin* (d-α-aminobenzylpenicillin) (Figure 5.7) showed superior therapeutic efficacy. It inhibits all bacteria susceptible to penicillin G, and most strains of *E. coli, Proteus mirabilis,* and *Haemophilus influenzae*. However, it is inactive against most strains of *Klebsiella, Enterobacter, Proteus,* and totally inactive against *Pseudomonas*. It was later clarified that the lack of activity against these strains is related to the presence of inducible β-lactamases. Ampicillin is also sensitive to the action of staphylococcal penicillinase and thus is inactive against *Staphylococcus aureus* strains that produce this enzyme.

Because ampicillin is fairly resistant to acid degradation, it is used orally. However, its oral absorption is not entirely satisfactory and many derivatives have been prepared to improve this aspect. Among these is *cyclacillin* (Figure 5.7), one of the best orally absorbed penicillins. On the other hand, cyclacillin, with a quaternary carbon atom in the position α to the amide, is less active than ampicillin against gram-negative strains. Derivatives with spectra of antibacterial activity very similar to ampicillin are *epicillin* and *amoxycillin* (Figure 5.7). The latter is almost completely absorbed orally and therefore, besides showing higher efficacy, is responsible for a lower incidence of intestinal disorders.

A different approach toward improvement of the oral absorption of ampicillin involves the preparation of lipophilic derivatives inactive *in vitro* but easily hydrolyzed in the body to give the active free antibiotic. As the simple esters are not easily hydrolyzed by the serum esterases because of the proximity of the bulky thiazole ring, double esters of the type -CO-O-CH_2-O-CO-R have been developed. The enzymatic hydrolysis of the second ester group unmasks a hemiacetal group, which

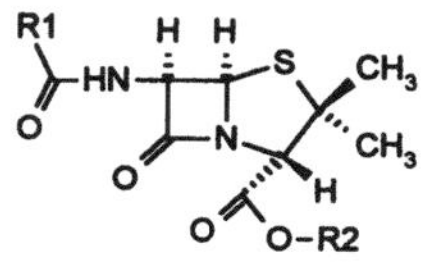

R1	R2	Name	MIC (µg/mL)					Oral absorption *
			S.aureus	*S.pyogenes*	*E.coli*	*P.mirabilis*	*H.influenzae*	
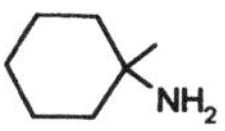	H	Cyclacillin	0.3	0.2	8	4	6	750
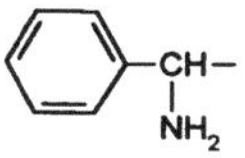	H	Ampicillin	0.06	0.05	2	2	0.5	250
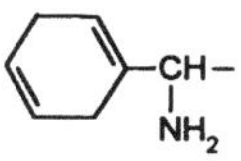	H	Epicillin	0.2	0.004	2	2	2	200
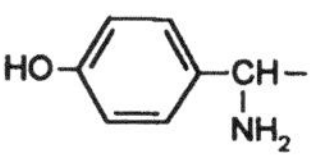	H	Amoxycillin	0.1	0.01	4	4	2	650
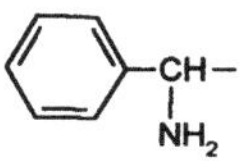	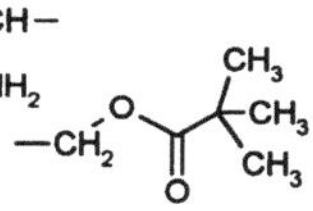	Pivampicillin	hydrolyzed to ampicillin					750

* peak serum levels in man, percent of those observed with penicillin G

Figure 5.7. Orally absorbed penicillins, active against *E. coli* and other enterobacteria.

spontaneously hydrolyzes. Among these, the most widely used is *pivampicillin* (Figure 5.7), an ampicillin ester that is well absorbed and rapidly hydrolyzed. Other esters such as *talampicillin* and *bacampicillin* have similar properties.

5.2.7. Anti-*Pseudomonas* and Anti-*Proteus* Penicillins

The increased activity against *E. coli* obtained by introducing weakly polar groups in the position α to the chain amide led to the preparation of molecules with stronger polar groups in this position for their poten-

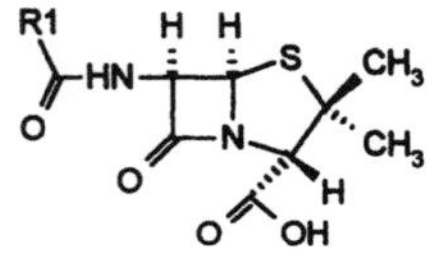

R1	Name	MIC (μg/mL)			
		E.coli	*P.vulgaris*	*P.mirabilis*	*P.aeruginosa*
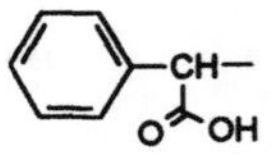	Carbenicillin	8	2	1	62
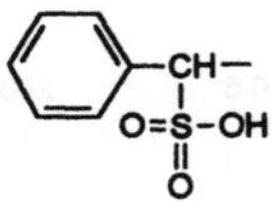	Sulbenicillin	12.5	25	-	31
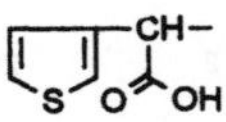	Ticarcillin	4	4	2	31
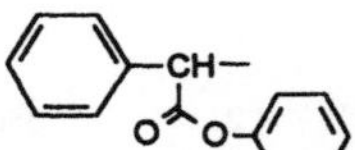	Carfecillin	hydrolyzed to carbenicillin			
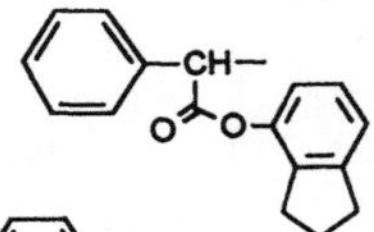	Carindacillin	hydrolyzed to carbenicillin			
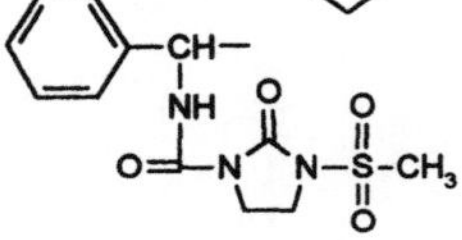	Mezlocillin	2	8	1	16
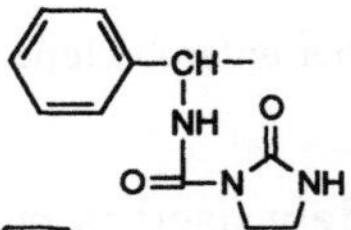	Azlocillin	8	16	2	1
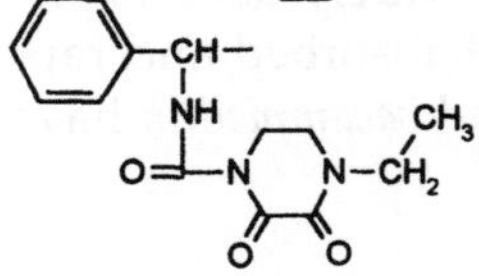	Piperacillin	1	4	0.5	4

Figure 5.8. Penicillins active against *Pseudomonas* infections.

tial activity against *Pseudomonas* or *Proteus* strains. The experiment was successful, as products with these structural moieties are insensitive to β-lactamases produced by these strains. Among the many products prepared, *carbenicillin, sulbenicillin,* and *ticarcillin* were sufficiently active to be introduced into clinical use (Figure 5.8). It must be noted, however, that these products show a decreased activity against gram-positives, especially enterococci, because of a lower affinity to PBPs.

The doses needed are often very high, up to 20 or even 40 g daily in very severe cases, although much lower doses may suffice for urinary tract infections or against more susceptible organisms. These products are not absorbed orally.

Esters readily hydrolyzed in the body fluids, such as *carfecillin* and *carindacillin,* can be administered orally but generate low blood levels. Later, a series of ureido derivatives of ampicillin, i.e., *mezlocillin, azlocillin,* and *piperacillin* (Figure 5.8), proved to be active against *Pseudomonas* and *Proteus*. Their MICs are quite low as they possess higher affinity for PBPs. While being less sensitive to β-lactamases than carbenicillin, a drawback is their low bactericidal activity.

5.2.8. Modification of the Penicillin Nucleus

None of the modifications described below resulted in increased antibacterial activity. Other modifications, which have generated interesting activities, are discussed in Section 5.2.14.

1. Modification of the carboxyl group
    a. Esterification brings about almost complete inactivation. However, as mentioned earlier, some esters are well absorbed orally and are rapidly hydrolyzed in body fluids, thus releasing the active penicillin.
    b. Activity is retained after conversion of the carboxyl to a thioacid or to an amide.
    c. Reduction to a hydroxyl group results in inactivation.
2. Modification of the nucleus
    a. Opening of either of the rings causes inactivation.
    b. Inversion of configuration at one of the three centers of asymmetry causes a great reduction in activity.
3. Other substitutions
    a. The presence of the two methyl groups is not required for activity.

b. Substituting a methyl or a methoxyl group for the hydrogen on carbon 6 results in decreased activity (however, the methoxyl group contributes to the resistance to β-lactamases). If the group is heavier, the activity disappears.
c. Oxidation of the sulfur atom to sulfoxide reduces the activity.

5.2.9. Natural Cephalosporins and 7-Aminocephalosporanic Acid

All cephalosporins originate from cephalosporin C (Figure 5.9), an antibiotic isolated in the 1950s from cultures of *Cephalosporium*, a mold noted for its ability to produce penicillin N (see Figure 5.2). *Cephalosporium* also produces a third antibiotic possessing a steroid structure, cephalosporin P, which has no therapeutic importance.

Not very effective, poorly absorbed orally, cephalosporin C attracted the attention of researchers because, although structurally related to the penicillins, it was active against penicillinase-producing *Staphylococcus aureus* and was more active against gram-negative bacteria. It was felt to be an interesting starting material for the preparation of semisynthetic derivatives. The experience with penicillins directed the research toward the preparation of 7-aminocephalosporanic acid (7-ACA) (Figure 5.9), obtained from cephalosporin C by a rather complex chemical reaction. In contrast to penicillins, an enzymatic method of hydrolyzing the amide in the chain of cephalosporin C became available only later. It was then demonstrated that this is related to the structure of the lateral chain, as semisynthetic cephalosporins with different chains are easily hydrolyzed enzymatically.

With 7-ACA as starting material, a large research effort has yielded an enormous number of semisynthetic cephalosporins, several of which are now in clinical use. Cephalosporins, unlike penicillins, can be modified not only at the side chains but also in position 3 of the nucleus,

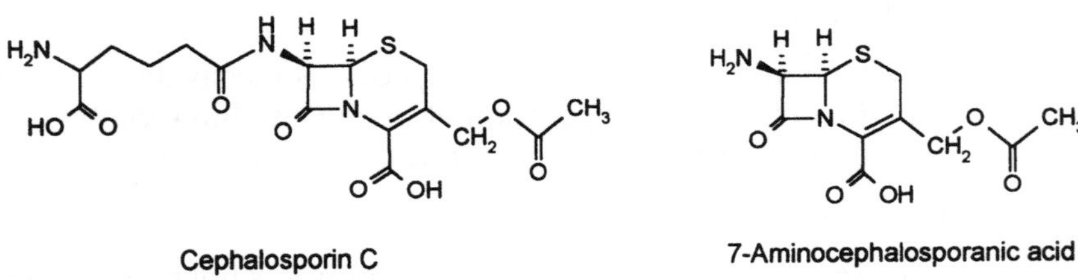

Figure 5.9. Structures of cephalosporin C and 7-aminocephalosporanic acid.

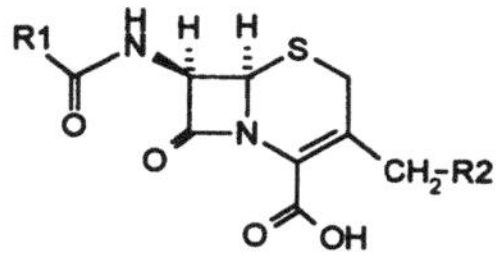

R1	R2	Name	MIC (μg/mL)					
			S.aureus	*S.aureus* (penicillinase producer)	*E.coli*	*K.pneumoniae*	*P.mirabilis*	*P.vulgaris*
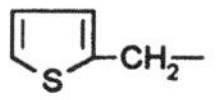	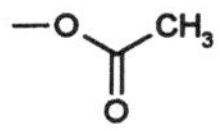	Cephalothin	0.04	0.2	6	4	8	>100
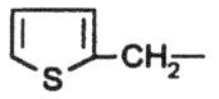	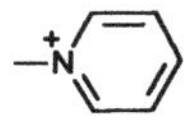	Cephaloridine	0.04	0.2	6	4	4	>100
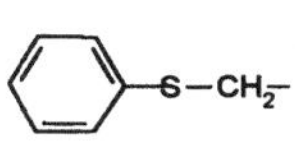	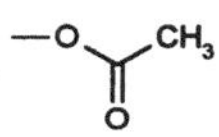	Cephapirin	0.2	0.4	12	6	12	>100
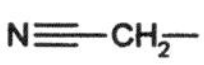	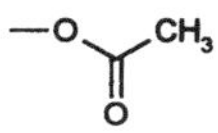	Cephacetrile	0.4	2	4	6	12	50
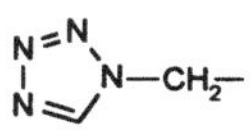	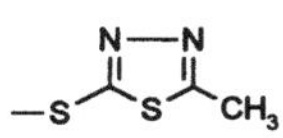	Cefazolin	0.1	0.8	1	4	2	>100

Figure 5.10. First-generation cephalosporins.

where the acetoxyl group can be easily eliminated or substituted without loss of activity.

5.2.10. First-Generation Cephalosporins

The first cephalosporin introduced into medical practice was *cephalothin,* in which a thienylacetic acid substitutes for aminoadipic acid in the side chain (Figure 5.10). Cephalothin is active against staphylococci both susceptible and resistant to penicillin, and against *Neisseria* and most *E. coli, Salmonella,* and *Proteus mirabilis* strains. The limitations to its clinical use, which have prompted further research in the cephalosporin field, may be summarized as follows:

1. It is not absorbed orally.
2. Intramuscular injections are painful and intravenous administration may produce phlebitis.

3. It is inactive against *Pseudomonas aeruginosa, Proteus* indole-positive, *Serratia marcescens, Enterobacter,* and *Bacteroides fragilis* strains.
4. It is rapidly eliminated (it is practically undetectable in serum 4 h after administration), thus requiring frequent doses.

In general, a wider range of structural modifications appear to be compatible with good antimicrobial activity in cephalosporins than in penicillins. However, it is more difficult to rationalize the relationship between the structure of the chemical substituents and the biological properties of the cephalosporins, in part because there are two modifiable sites in the molecule and the ensuing biological properties are the results of the combined effects rather than of the sum of the substitutions.

Since the acetyl group in position 3 is readily hydrolyzed in the body to yield the less active hydroxyl derivatives, substitutions at this position improve the pharmacokinetic properties (Figure 5.10). The substitution of the acetoxy group of cephalothin with a pyridine ring has produced *cephaloridine.* It possesses an antibacterial activity similar to that of cephalothin but injections are less painful and produce higher serum levels. However, high doses may be nephrotoxic, a toxicity seldom found with other cephalosporins. Properties similar to those of cephalothin are shown by *cephapirin* and *cephacetrile,* both of which are modified only in the amide side chain. Again, a better tolerability after intramuscular injections is their major characteristic. Cephapirin seems less active on gram-negative bacteria, whereas cephacetrile is somewhat more active than cephalothin against *E. coli.*

Cefazolin, with a mercapto thiadiazole at position 3 and a tetrazole in the amide chain, has about the same activity as cephalothin on gram-positives (although it is slightly less resistant to *S. aureus* β-lactamases) but is somewhat more active against *E. coli, K. pneumoniae,* and *Salmonella.* In addition, it is better tolerated after intramuscular injections and can be administered at longer intervals, having a longer serum half-life.

5.2.11. Orally Active Cephalosporins

The synthesis of new cephalosporins obviously took advantage of the knowledge accumulated during years of penicillin studies. The phenylglycine side chain of ampicillin was shown to be an excellent moiety for cephalosporins also. Almost all the orally active cephalosporins used at present (Figure 5.11) possess either this side chain or small variants of it, such as cephradine, an analog of epicillin, and cefadroxil and cefatrizine, analogues of amoxycillin.

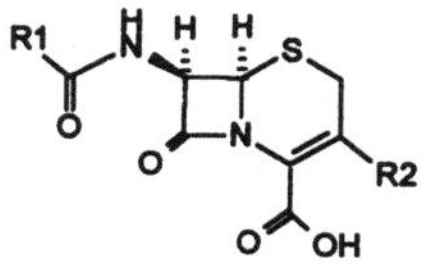

R1	R2	Name	MIC (µg/mL) S.aureus	MIC (µg/mL) E.coli	MIC (µg/mL) H.influenzae	MIC (µg/mL) P.mirabilis	Oral absorption (µg/mL) *	(hours) **
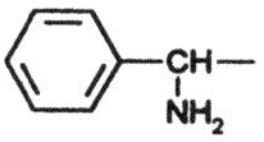	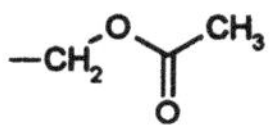	Cephaloglycine	2	4	-	6	1	-
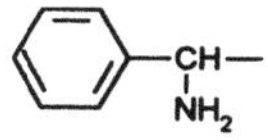	$-CH_3$	Cephalexin	4	16	4	8	20	0.8
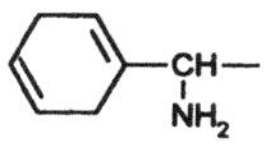	$-CH_3$	Cephradine	4	16	16	16	15	-
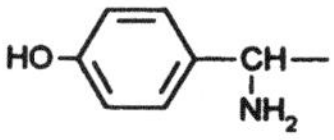	$-CH_3$	Cefadroxil	2	8	16	16	20	1.2
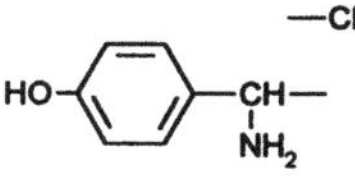	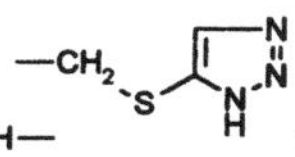	Cefatrizine	0.4	4	4	2	6	2.4
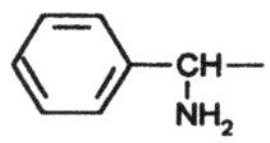	—Cl	Cefaclor	2	1	2	1	15	0.8
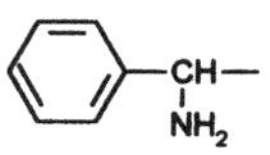	$-O-CH_3$	Cefroxadine	16	4	4	8	-	-
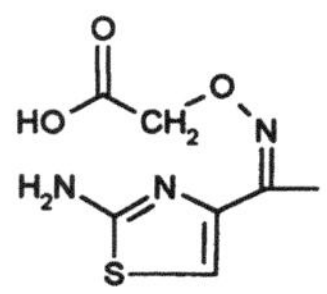	$-HC{=}CH_2$	Cefixime	4	0.25	0.06	0.06	-	3.5

* peak serum level in man after 500 mg os ** half-life in serum

Figure 5.11. Orally active cephalosporins.

The first orally active cephalosporin, *cephaloglycine*, has the natural substituent, acetoxymethyl, at position 3. Its spectrum of antibacterial activity is similar to that of cephalothin, and includes most gram-positive bacteria (with the exception, common to most cephalosporins, of *Enterococcus faecalis*), and *Neisseria*, *E. coli*, and *Proteus mirabilis*. However, higher concentrations are usually needed for inhibition. Today, it has been practically abandoned because it is rapidly metabolized to the poorly active desacetyl derivative.

Cephalexin and *cephradine* are more stable and better absorbed, which compensates amply for their lower *in vitro* activity. *Cefadroxil* is similar to cephalexin with its spectrum of antibacterial action and has a longer serum half-life.

A combination of good absorption and greater activity against some gram-negative bacteria was sought in the derivatives synthesized later, mainly by varying the substituent at position 3. *Cefatrizine* belongs to this type of cephalosporin. Its absorption after oral administration is in fact rather poor; however, it is generally more active than the previously available derivatives against several strains including *Haemophilus influenzae*. A further improvement was obtained by synthesizing derivatives with small lipophilic substituents in position 3. In *cefaclor* this substituent is chlorine. Cefaclor is at least equivalent to cephalexin in activity against gram-positive organisms and is more active against gram-negatives, including *H. influenzae* and *N. gonorrhoeae*. *Cefroxadine*, where the substituent in position 3 is methoxyl, is somewhat less active. *Cefixime*, which bears a vinyl group in position 3 and a complex chain in position 7 (derived from the experience of the injectable third-generation cephalosporins), is very active against almost all gram-negative bacteria.

A different approach, derived from the experience on the easily hydrolyzable esters of penicillins, has been adopted for preparation of *axetil-cefuroxime*, which possesses the properties of the parent cefuroxime (see Section 5.2.12), and, in addition, is well absorbed orally.

5.2.12. Second- and Third-Generation Injectable Cephalosporins

These cephalosporins are characterized by their enhanced antibacterial activity, obtained mainly by the choice of a suitable substituent at the amide chain combined with a substituent in position 3, compatible with good pharmacokinetics (Figure 5.12). The antibacterial spectra of *cefamandole* and *cefuroxime* (second generation) include indole-positive *Proteus* and *Enterobacter* species, and *H. influenzae*. Cefamandole maintains good activity against gram-positive species and is more active than first-generation cephalosporins against *E. coli*, *K. pneumoniae*, and *P. mirabilis*.

The prototype of third-generation cephalosporins is *cefotaxime*, characterized by an acyl chain, which bears in α a methoxyimino group (the *syn* isomer, unlike cefuroxime, that has the *anti* isomer) and in β an aminothiazole. The presence of the substituent in α protects the mole-

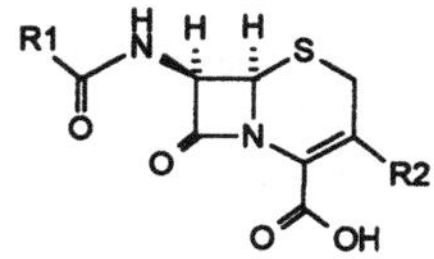

MIC (μg/mL)

R1	R2	Name	*S. aureus*	*S. pyogenes*	*E. coli*	*P. mirabilis*	*P. vulgaris*	*P. aeruginosa*
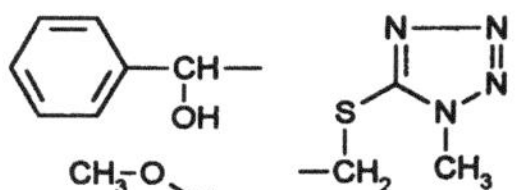		Cefamandole	0.06	0.04	0.2	1	>100	>100
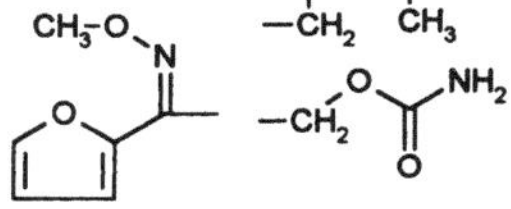		Cefuroxime	0.2	0.01	4	2	>100	>100
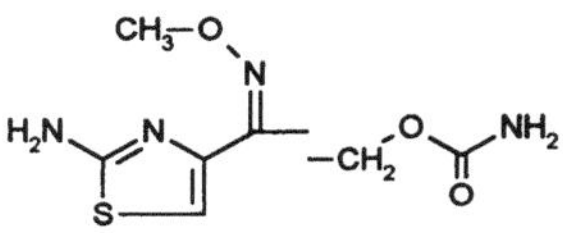		Cefotaxime	1	0.02	0.1	0.06	0.1	16
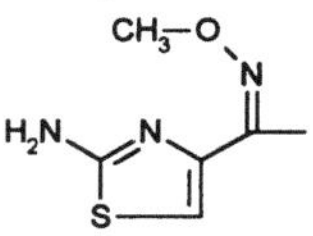	H	Ceftizoxime	1	0.1	0.1	0.03	0.06	16
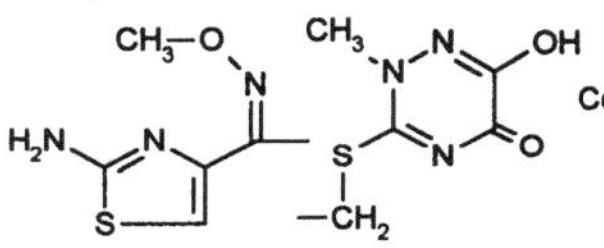		Ceftriaxone	1	-	0.06	0.01	0.01	8
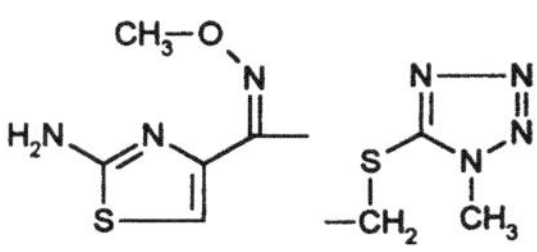		Cefmenoxime	0.5	-	0.1	0.1	0.1	16
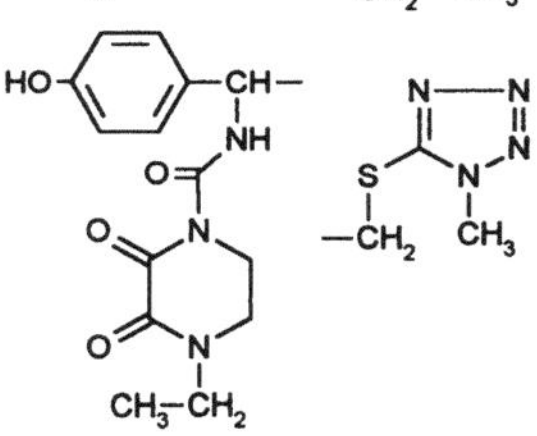		Cefoperazone	8	0.1	0.1	1	16	4

Figure 5.12. Second- and third-generation injectable cephalosporins.

cule from the attack of β-lactamases, and thus cefotaxime combines an excellent activity against enterobacteria with an acceptable activity against gram-positives (with the common exception of enterococci) and a fair activity against *Pseudomonas*. Aminothiazole enhances the affinity toward the PBPs, particularly PBP1 and PBP3 of gram-negatives. However, the acetyl group in 3 of cefotaxime is rapidly hydrolyzed in blood. Thus, several derivatives have been proposed, with the same chain in position 7 but with different substituents in position 3. Among these, *ceftizoxime*, *ceftriaxone*, and *cefmenoxime* possess antibacterial activities quite similar to those of cefotaxime, but display better pharmacokinetic properties.

Cefoperazone derives from the experience of ureido-penicillins. It possesses a good spectrum of activity as it is resistant to the action of penicillinases of staphylococci and of some gram-negatives.

5.2.13. Anti-*Pseudomonas* Cephalosporins

Second- and third-generation cephalosporins, although having a broad spectrum of activity against gram-negatives, show poor activity against *Pseudomonas*. Analogously to what was observed in penicillins, it became evident that the presence of a strongly acidic group in the position α to the amidic carboxyl increased the antibiotic's activity against these microorganisms (Figure 5.13). The typical example of this class is *cefsulodin*, whose anti-*Pseudomonas* activity is, however, still fairly poor.

A more active product, *ceftazidime*, was obtained by keeping the oxyimino group as substituent in position α of the side chain, but substituting the methyl group with a chain terminating with a carboxyl group. The presence of a pyridinium group in position 3 enhances the activity.

A similar approach was applied in the design of *cefpirome* and *cefepime*, in which the anti-*Pseudomonas* activity is based on the presence, in addition to the methoxyimino group of the side chain, of a quaternary ammonium within the substituent in position 3.

5.2.14. Nonclassic β-Lactams

5.2.14.1. Amidino Penicillins

The initial concept, based on a large amount of data, that an amide in position 6 of aminopenicillanic acid is an absolute prerequisite for activity, was later contradicted by the observation that amidino derivatives of 6-aminopenicillanic acid possess antibacterial activity. The most interesting derivative of this series is *amdinocillin*, previously called

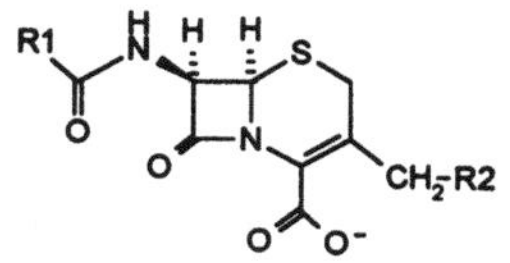

MIC (μg/mL)

R1	R2	Name	S.aureus	E.coli	P.mirabilis	P.vulgaris	P.aeruginosa
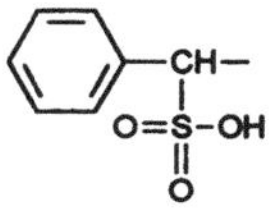	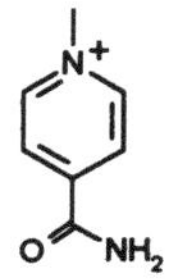	Cefsulodin	4	31	62	125	0.5
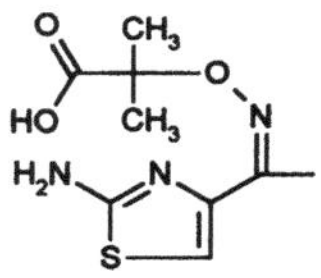	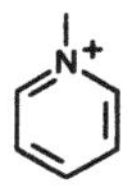	Ceftazidime	4	0.06	0.06	0.06	1
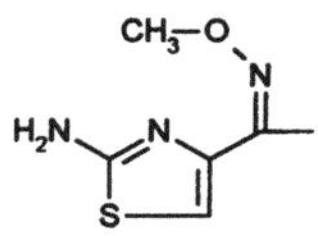	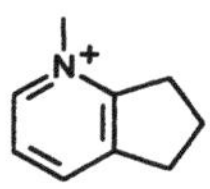	Cefpirome	0.5	0.06	0.1	0.1	2

Figure 5.13. Cephalosporins active against *Pseudomonas*.

mecillinam (Figure 5.14), which was shown in clinical trials to be effective as well as its ester *pivmecillinam*, which is active orally. The antibacterial activities reported in Figure 5.14 show that amdinocillin's spectrum of action substantially differs from that of the penicillins. This is because its site of action is not the same as that of classic penicillins. Amdinocillin mainly interferes with PBP2 of *E. coli* and has a poor affinity to the PBPs of gram-positives.

5.2.14.2. Cephamycins and Other Methoxy-Substituted β-Lactams

Cephamycins are naturally occurring β-lactam antibiotics first isolated in 1971 from *Streptomyces*. Structurally, they are cephalosporins that bear a methoxy group in position 7α (Figure 5.15). Among the several natural ones, *cephamycin C* was noted as it showed higher activity against gram-negative than gram-positive bacteria. It was later verified that the presence of the methoxy group confers a marked stability toward the action of various β-lactamases.

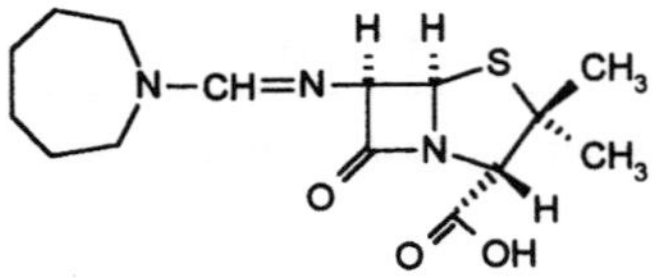

	MIC (µg/mL)
S.aureus	5
E.faecalis	>100
E.coli	0.02
H.influenzae	16
P.mirabilis	0.1
P.vulgaris	0.2
P.aeruginosa	>128
S.typhimirium	0.1

Figure 5.14. Structure and *in vitro* activity of amdinocillin.

Like cephalosporin C, cephamycin C was thus considered a suitable starting material for the synthesis of more active derivatives. Two of these, *cefoxitin* and *cefmetazole* (Figure 5.15), have been introduced into clinical practice. In general, they are similar to second-generation cephalosporins. They are fairly active against gram-positives, but very active against *E. coli*, *Proteus*, and anaerobes, such as *Bacteroides* and clostridia.

Cefotetan was introduced in clinical practice more recently. It is more active than cefoxitin against various gram-negative aerobes, is less sensitive to the action of β-lactamases, and possesses a longer half-life (3.5 instead of 0.8 h).

The protection shown by the methoxy group toward inactivation from β-lactamases has suggested the synthesis of other β-lactams containing this group. Among the penicillins, *temocillin*, structurally a 6α-methoxy-ticarcillin, showed greater stability toward β-lactamases and consequently greater activity against gram-negative bacteria. However, temocillin is inactive against gram-positives and is poorly active against *Pseudomonas*.

In the cephalosporin ring an oxygen or a methylene can substitute for the sulfur atom without substantial loss of activity. This observation opened the way to the synthesis of a completely new series of derivatives. The most important of these products, *moxalactam* (Figure 5.15), has, like cephamycins, a methoxy group in position 7, and shows an activity similar to that of third-generation cephalosporins.

5.2.14.3. Carbapenems

These antibiotics are characterized by a β-lactam ring condensed with a pyrroline ring and thus differ from penicillins in the substitution

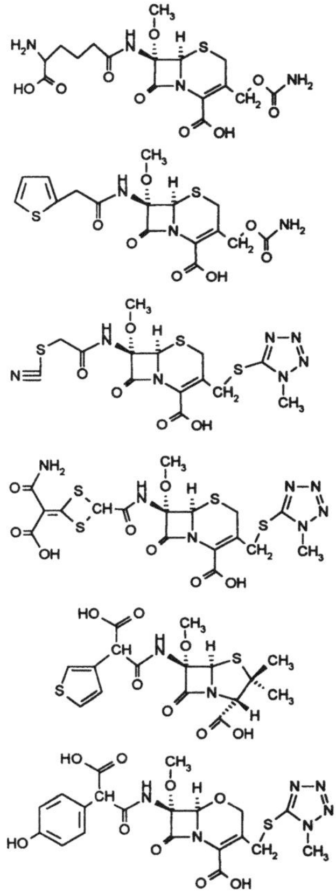

Name	MIC (μg/mL)					
	S.aureus	E.coli	P.mirabilis	P.vulgaris	B.fragilis	P.aeruginosa
Cephamycin C	160	16	-	-	-	-
Cefoxitin	2	2	1	8	2	>100
Cefmetazole	0.2	0.5	0.5	2	2	>100
Cefotetan	4	1	0.5	2	4	32
Temocillin	>100	4	1	1	>100	>100
Moxalactam	4	0.1	0.01	0.1	0.5	4

Figure 5.15. Methoxy-substituted β-lactam antibiotics.

of a methylene for the sulfur atom. In addition, instead of the amide chain in position 6 they have a short aliphatic chain (Figure 5.16). Several carbapenems have been isolated from actinomycete strains. The most important is *thienamycin,* whose name derives from the presence of a substituent at the pentene ring initiating with a sulfur atom. Thienamycin has a considerably broad antibacterial spectrum comprising most gram-positives and gram-negatives, including *Pseudomonas aeruginosa*. It shows a marked affinity for PBP 2 of *E. coli* and *P. aeruginosa* and a fair affinity for PBP 1, 4, 5, and 6. In addition, it is insensitive to most β-lactamases. However, thienamycin is markedly unstable, both chem-

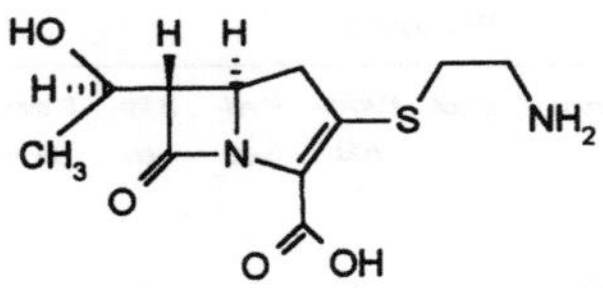

Thienamycin

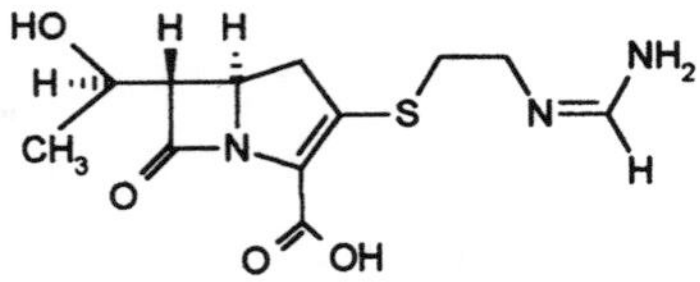

Imipenem

	MIC (μg/mL)
S.aureus	5
S.pyogenes	0.1
E.faecalis	>100
E.coli	0.02
P.mirabilis	0.1
P.aeruginosa	>128
B.fragilis	0.2

Figure 5.16. Structures of carbapenems and *in vitro* activity of imipenem.

ically and metabolically. Different derivatives have been synthesized and, among them, *imipenem* (Figure 5.16) proved to be quite stable and active. It was then introduced into clinical practice in combination with *cilastatin,* an inhibitor of peptidases, to prevent the metabolic inactivation of the antibiotic by these enzymes. It is noteworthy that the resistance to β-lactamases of thienamycin and imipenem is specifically related to the configuration of the side chain hydroxyl group. It is also noteworthy that, unlike penicillins, the hydrogen atoms in positions 5 and 6 are *trans*.

5.2.14.4. Monobactams

A research program aimed at isolating new β-lactams produced by less common microorganisms, such as gram-negatives living in the soil, has yielded a new class of β-lactam antibiotics characterized by the presence of the β-lactam ring not condensed with other rings. These monocyclic β-lactams are called monobactams (Figure 5.17).

The first monobactams isolated, *sulfazecin* and *isosulfazecin*, are characterized by a sulfonic group as substituent at the lactam nitrogen and by a dipeptide in the position α to the carbonyl. Their activity is quite

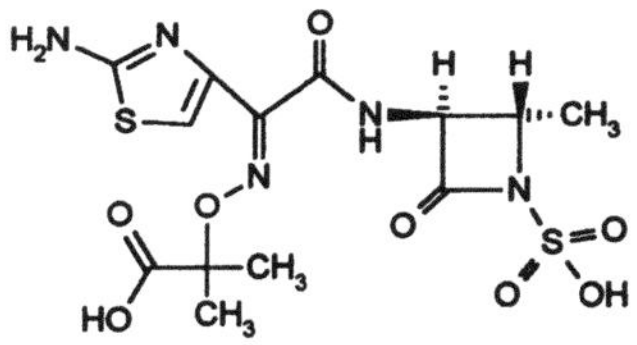
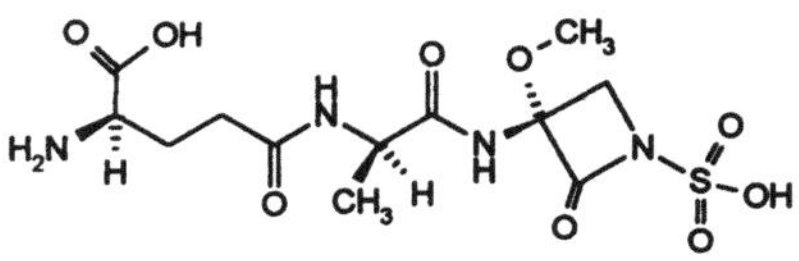
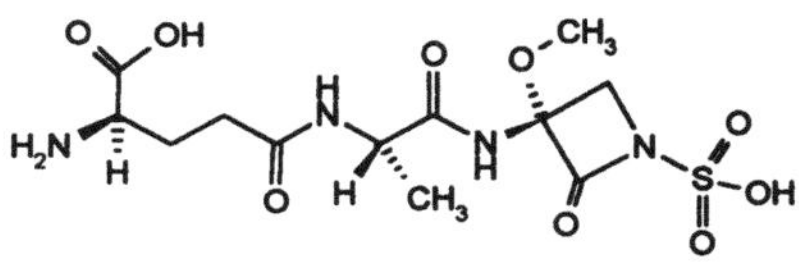

	MIC (μg/mL)				
	S.aureus	*E.coli*	*P.mirabilis*	*P.vulgaris*	*P.aeruginosa*
Aztreonam	>100	0.1	0.1	0.1	4
Sulfazecin	200	6.2	3.1	-	100
Isosulfazecin	200	50	200	-	100

Figure 5.17. Structures and *in vitro* activities of monobactams.

poor against gram-positive bacteria, but fair against gram-negative ones, and they have been used as starting materials for the synthesis of derivatives. Among the numerous products synthesized, *aztreonam* has found clinical use. It is insensitive to β-lactamases and is particularly active against gram-negative aerobes, including some strains of *P. aeruginosa*. Its use is mainly justified by the lower toxicity with respect to other antibiotics possessing the same spectrum of action and to the apparent lack of cross-allergenicity with penicillins.

5.2.14.5. β-Lactamase Inhibitors

Methicillin and other penicillins resistant to staphylococcal β-lactamases act also as inhibitors of the β-lactamases, and in theory they could be used in combination with cephalosporins that are inactivated by these enzymes, to increase their activity or their spectrum of action. In practice, their use is limited by the fact that they hardly penetrate the outer membrane of gram-negatives. A more effective inhibitor is *clavulanic acid*, isolated from *Streptomyces* strains. This molecule is com-

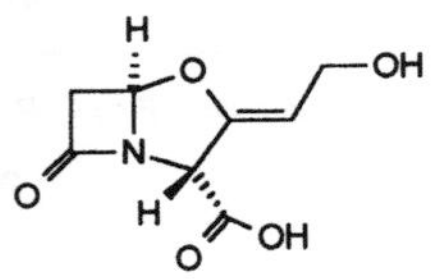

Clavulanic acid

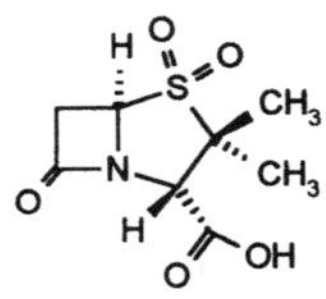

Sulbactam

Figure 5.18. Structures of some β-lactamase inhibitors.

posed by a β-lactam ring condensed with an oxazoline giving rise to a penicillin nucleus with an oxygen substituting for the sulfur atom (Figure 5.18). Clavulanic acid has an antibacterial spectrum quite broad but with marginal activity. It penetrates well into the gram-negatives and it binds covalently to various β-lactamases inactivating them. Thus, it can be considered a suicide inhibitor of β-lactamases of staphylococci and several gram-negatives such as *Proteus, Escherichia,* and *Haemophilus,* but not of *Pseudomonas* or *Enterobacter*. It is used in combination with amoxycillin (the combination is called "augmentin") to treat infections from bacteria resistant to amoxycillin.

Another inhibitor of β-lactamases is *sulbactam,* in which the sulfur of thiazoline is oxidized to sulfone (Figure 5.18). It shows an activity similar to that of clavulanic acid but it has a much lower ability to penetrate the outer membrane of gram-negatives. It is used parenterally in combination with ampicillin.

5.2.15. Considerations on the Structure-Activity Relationship in β-Lactams

Some general conclusions can be drawn from the available data concerning the structural characteristics responsible for these antibiotics' intrinsic activity. The basic element is the tension of the β-lactam structure, which creates the possibility to covalently react with the PBPs. In penicillins, the reactivity is increased by the tension of the β-lactam ring fused with the pentatomic one, a fact that increases the pyramidal nature of nitrogen and brings about a decreased delocalization of the nonpaired electrons. In cephalosporins, the tension is lower but the lone pair of nitrogen interacts with the double bond in position 3—4, thus decreasing the stability of the amide function. A similar influence can be hypothesized for the double bond in carbapenems. In monobactams, the tension is lower because the second ring is lacking, but the reactivity is increased by the sulfonic residue bound to nitrogen.

It is clear that higher reactivity implies less stability, especially in acidic mediums. For practical use, it is necessary to find a compromise between stability and activity, especially for the derivatives designed for oral use. This can be obtained by introducing electron-withdrawing substituents in the chain at position 6 of penicillins or position 7 of cephalosporins, which decrease the reactivity of the β-lactam carbonyl. Nevertheless, it must be taken into consideration that the intrinsic activity is mainly a result of, in addition to the lactam reactivity, the general structure of the molecule, which determines the degree of affinity toward one or more of the various PBPs.

Another consequence of the higher reactivity is an increased susceptibility toward β-lactamases. However, in this case also, the general structure of the molecule determines the degree of affinity toward each of the various enzymes. As has been shown, the presence of bulky groups at particular positions decreases this affinity and consequently makes the molecule an unsuitable substrate for enzymatic reaction.

5.2.16. Problems in the Use of β-Lactam Antibiotics

5.2.16.1. Spectrum of Activity and Resistance

Following the introduction of the new β-lactam antibiotics into clinical practice, the whole class can now be considered active against almost all the bacteria responsible for the most common infectious diseases, with the exception of mycobacteria and, obviously, of mycoplasma, which lack cell walls. However, there are microorganisms against which the activity of β-lactam antibiotics is unsatisfactory. For example, enterococci are only moderately susceptible to penicillins and almost totally insusceptible to most cephalosporins. In addition, the susceptibility of the different gram-negative organisms, such as *Enterobacter*, *Klebsiella*, *Serratia*, and indole-positive *Proteus*, is variable from strain to strain. Similarly, various species of *Pseudomonas* are generally less sensitive and sometimes insensitive.

From a bacteriological point of view, however, the main problem is the high frequency of resistant strains. Particularly serious is the situation with staphylococci and some other gram-positives because of the presence of the so-called methicillin-resistants, which, possessing a particular type of PBP, are insusceptible to the β-lactam antibiotics. These resistant strains are especially prevalent in hospitals and they pose a serious therapeutic threat as they often bear genes for resistance toward other classes of antibiotics. The other common form of resistance, the production of β-lactamases, can be overcome in part, as previously described, by the use of specific derivatives insusceptible to these enzymes, or, alternatively, by the use of a combination of an active product and an inhibitor of β-lactamases, such as clavulanic acid or sulbactam. It appears more difficult to inhibit those strains, in particular *Pseudomonas*, in which the penetration of the antibiotic is reduced by a mutation of the porins.

Another phenomenon of potential clinical importance is that of "tolerance." Tolerant strains are inhibited by a "normal" low concentration of antibiotic but show a much higher minimal bactericidal concentration than nontolerant strains. Penicillins function as bacteriostatic rather than bactericidal antibiotics toward these strains. Since with the usual dosages of penicillins the effectiveness is linked to their bactericidal action, the infections sustained by tolerant strains may not be cured efficiently.

5.2.16.2. Tolerability and Secondary Effects

In general, β-lactam antibiotics are well tolerated because of their specific mechanism of action. They show some secondary effects that are common to many drugs, such as gastric troubles or nausea when administered orally, and, in some cases, low local tolerability when administered parenterally. As frequently occurs with broad-spectrum antibiotics, there is the risk of superinfections or selection of resistant strains at the intestinal level, particularly for the derivatives that are administered orally or for those that possess a high biliary excretion.

The most specific secondary effects are immunological reactions and some hematological effects.

Hypersensitivity. Allergic skin reactions are the most common manifestation of hypersensitivity to β-lactam antibiotics. The incidence of such reactions observed with penicillins is 7–10% of patients. It is difficult to predict these reactions as there is a poor correlation between the most used hypersensitivity test (cutaneous reaction toward polylysylpenicillin) and the allergic manifestation. The incidence of cutaneous reactions to cephalosporins is lower (1–3%).

It has been shown that there is a partial cross-immunoreactivity between penicillins and cephalosporins. This seems not to be the case with some nonclassical β-lactams and aztreonam. A different type of immunological reaction is the so-called "flu syndrome," which manifests itself with fever, myalgia, and atralgia. It occurs with a frequency that varies according to the derivative and seems correlated to high dosages and prolonged treatments.

Very serious but fortunately very rare are the episodes of anaphylactic shock, whose incidence is estimated as one case over 50,000–100,000 dose-treatments.

Hematological reactions. In addition to causing some reactions of minor importance, some β-lactams, particularly penicillins with a carboxyl group, can interfere with platelet aggregation, causing hemorrhages. A more serious secondary effect is inhibition of the synthesis of prothrombin, which is caused by some cephalosporins such as cefamandole and cefoperazone, and by other derivatives such as moxalactam and cefotetan. This effect has been hypothesized to be caused by lack of vitamin K, resulting from the suppression of the intestinal bacterial producers of this vitamin. However, it is today believed that these products may have a direct inhibitory effect on the synthesis of prothrombin in the liver. In fact, it has to be noted that all these derivatives bear a methyltetrazole in position 3 and that this molecule, when administered alone, shows an inhibitory effect on the prothrombin synthesis in rats.

5.3. Tetracyclines

Tetracyclines form a small family of antibiotics, whose name derives from their chemical structure of four linearly condensed rings. From this point of view they are related to the wider family of anthracyclines, from which they differ by possessing a higher degree of reduction (only one ring of tetracyclines is aromatic) and different substituents (among them, a dimethylamino group). However, the most important difference is biological. In fact, tetracyclines inhibit protein synthesis at the ribosomal level, while anthracyclines usually interfere with DNA functions. The structure–activity relationship for the various tetracyclines described later indicates that variations in the upper periphery of the molecule (as represented in Figure 5.19) maintain the antibacterial activity, while variations in other regions cause inactivation.

5.3.1. Natural Tetracyclines

The introduction of tetracyclines in therapy began in 1948 with *chlortetracycline*, produced by *Streptomyces aureofaciens*. This was followed two years later by *oxytetracycline*, isolated from *Streptomyces rimosus*, and in 1952 by *tetracycline* (Figure 5.19). The latter was originally synthesized by catalytic hydrogenation of chlortetracycline and later by fermentation, first by growing *S. aureofaciens* in a medium without chlorine and later by fermentation of *Streptomyces* strains that were able to produce it even in chlorine-containing media.

All three compounds show complete cross-resistance and all have very similar spectra of activity that are particularly broad, encompassing not only gram-positive and gram-negative bacteria and mycoplasmas,

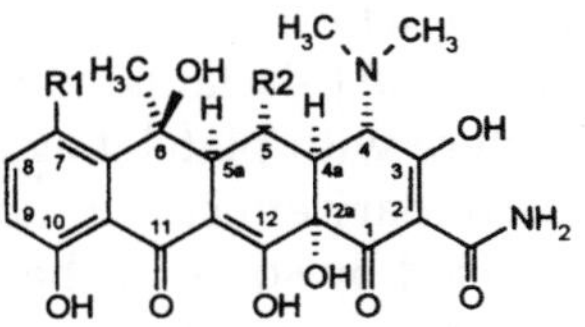

R1	R2	Name	*S.aureus*	*S.pneumoniae*	*E.faecalis*	*E.coli*	*K.pneumoniae*	*P.vulgaris*	*S.tiphymurium*	*P.aeruginosa*
			MIC (μg/mL)							
H	H	Tetracycline	0.4	0.15	1.2	0.7	0.6	4	2.3	25
Cl	H	Chlorotetracycline	0.3	0.1	1.4	0.4	0.3	4.6	1.2	14
H	OH	Oxytetracycline	0.6	0.3	2	1.2	0.6	3.1	1.6	25

Figure 5.19. Natural tetracyclines.

but also such intracellular microorganisms as *Rickettsia* and *Chlamydia*. They are, however, almost inactive against *Proteus* and *Pseudomonas*. In very high concentrations they also inhibit some protozoa. In spite of some claims, they have no effect against viruses, which is only to be expected from their mechanism of action.

The therapeutic effectiveness is approximately equal for all tetracyclines and all are well, albeit not completely, absorbed when given orally. Chlortetracycline is slightly more active *in vitro* against some bacteria, but not more effective in experimental infections, probably because of its instability at physiological pH values and lower blood levels. Tetracycline is the most frequently used for human therapy.

5.3.2. Semisynthetic Tetracyclines Derivatized at the Amide Group

The first attempts to chemically modify the tetracyclines had a very practical and limited objective: to obtain a derivative that would be soluble in water at neutral pH and could therefore be injected without causing pain or irritation. This objective was rapidly reached through the synthesis of *rolitetracycline* (Figure 5.20) and of the analogous compounds *limecycline* and *mepicycline*, obtained from tetracycline by reaction with formaldehyde and the appropriate amines. These products have the same activity as tetracycline and, in fact, are hydrolyzed to it in

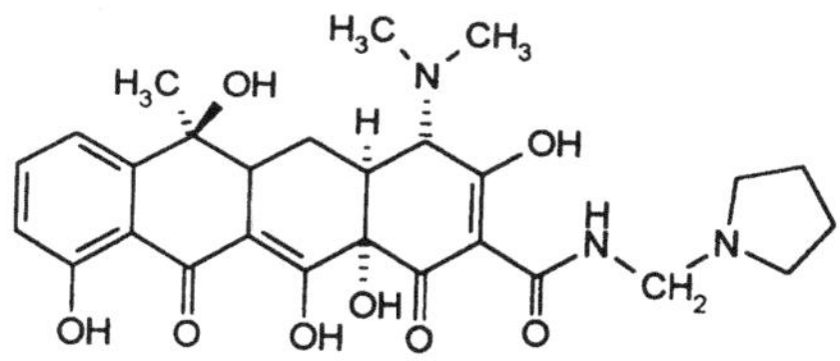

Figure 5.20. Structure of rolitetracycline, a soluble derivative of tetracycline.

aqueous solution. On the contrary, the amide derivatives with substituents that are not easily hydrolyzable are inactive.

5.3.3. Tetracyclines Produced Biosynthetically

The natural tetracyclines are rapidly converted into inactive derivatives by treatment with acid or base. Degradation products, made for the purpose of determining the structure, are also inactive. Therefore, it is not surprising that the first active derivatives, apart from the water-soluble ones mentioned previously, were not obtained by chemical modification but by modifying the biosynthetic pathway. The first interesting example was the preparation of *7-bromotetracycline,* which was obtained when bromide ions were substituted for chloride ions in the fermentation broth of *Streptomyces aureofaciens.* However, this product was less active than the natural tetracyclines and was never used.

A product that has been introduced into clinical use is *6-demethyl-7-chlortetracycline* (Figure 5.21), which was initially obtained by adding inhibitors of methylation, such as sulfonamides, to the fermentation broth, and later by isolating mutant strains with an inactive methylating enzyme, thereby producing the demethyl derivative instead of chlortetracycline. *Demethyltetracycline* and *demethyloxytetracycline* were obtained by the same method. *Demethylchlortetracycline* is very stable and has an antibacterial spectrum very similar to that of the original compounds. It is excreted more slowly, giving higher and more prolonged blood levels, thus allowing the use of lower dosages.

5.3.4. Semisynthetic Tetracyclines Modified in Positions 6 and 7

Comparisons of the activities of the natural tetracyclines and of different derivatives have shown that positions 5, 6, and 7 can have different substituents without any substantial loss of activity. Obviously these are not directly involved in the formation of the bond with the

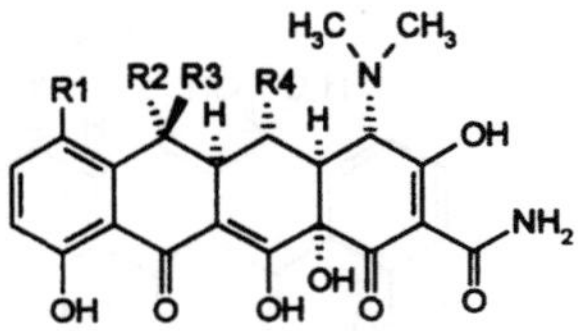

R1	R2	R3	R4	Name	MIC (μg/mL) *S.pyogenes*	*S.aureus* (tetracycline resistant)	*S.pneumoniae*	*E.coli*	*P.mirabilis*
Cl	H	OH	H	Demethylchloro-tetracycline	0.8	-	1	3	25
H	CH_2=		OH	Methacycline	0.8	-	1	3	25
H	CH_3–	H	OH	Doxycycline	0.4	-	0.04	1.6	50
$(CH_3)_2N$–	H	H	H	Minocycline	0.4	0.8	0.04	3.1	50

Figure 5.21. Modified tetracyclines.

biological receptor. Much attention was given by researchers to two reactions that allow modifications in these positions.

1. After halogenation with *N*-chlorosuccinamide in position 10a, a molecule of water can be removed from position 6 to give 6-methylene-tetracyclines. By this procedure, *methacycline* (6-desoxy-6-demethyl-6-methylene-oxytetracycline) was prepared. With this as starting material, it was possible to synthesize *doxycycline* (6-desoxy-oxytetracycline) (Figure 5.21). Both methacycline and doxycycline have been introduced into therapeutic use; the latter has the advantages of prolonged high blood levels and higher activity against some strains that are poorly susceptible to tetracycline.
2. Hydrogenolysis under specific conditions (with rhodium as catalyst) leads to loss of the hydroxyl in position 6. The activities of 6-desoxy-tetracycline and 6-desoxy-oxytetracycline so obtained are considerably decreased not because the hydroxyl was eliminated but because the reaction leads to the inversion of the configuration of the methyl group in position 6. In fact, doxycycline,

mentioned previously, also lacks the hydroxyl at position 6, but the methyl group is in the correct configuration and the activity remains as high as that of the natural oxytetracycline. This interpretation is confirmed by the activity of 6-desoxy-6-demethyltetracycline, which is equal to that of 6-demethyltetracycline.

Removal of the hydroxyl group confers great stability to tetracyclines. Therefore, 6-desoxy-6-demethyltetracycline is the ideal starting material for even more drastic chemical reactions carried out to obtain derivatives with substitutions at positions 6 and 7. Among these, the most important is *minocycline* (7-dimethylamino-6-demethyl-6-deoxytetracycline) (Figure 5.21), which is active both *in vitro* and *in vivo* against strains of *S. aureus*, *Streptococcus*, and *E. coli* (but not *Proteus* and *Pseudomonas*) resistant to tetracycline. Minocycline was introduced in therapy but the hope that it would represent a promising beginning leading to the synthesis of other tetracyclines without cross-resistance with the original ones has not materialized. Nowadays, only a very few laboratories are still working on new tetracyclines.

5.3.5. Unresolved Problems

5.3.5.1. Resistance

Strains resistant to tetracyclines can be produced in the laboratory by serial culture, but resistance develops slowly and is of the multistep type. In nature, however, there are many resistant strains, principally those that carry the R factor for transferable resistance. It should be pointed out that the low frequency of resistant mutants among the bacteria derived from the normal strains has the practical consequence that one seldom sees "conversion," i.e., the appearance of resistant bacteria, during treatment of patients who have infections sustained by susceptible bacteria. However, the widespread diffusion, especially in hospitals, of resistant strains is the cause of an increased number of infections refractory to tetracycline treatment. It was previously stated that minocycline is active against some mutants that are resistant to other tetracyclines: the clinical practice of the last few years has confirmed its efficacy especially against strains of staphylococci resistant to both methicillin and tetracycline.

5.3.5.2. Secondary Effects

Tetracyclines may cause hypersensitivity reactions and allergy, but with much lower frequency than the penicillins and the cephalosporins.

Gastrointestinal irritation is rather common and very high doses can be hepatotoxic. Because of their broad spectrum of action, partial absorption, and excretion in the bile, they can alter the intestinal flora to a great extent and consequently may give rise to superinfections, sometimes very serious, caused by yeasts or resistant strains of *S. aureus*.

Tetracyclines are not generally nephrotoxic, but worsening of preexisting renal pathologies has been observed. In particular, demethylchlortetracycline can induce injuries that manifest themselves as diabetes insipidus.

Other typical secondary effects, resulting from the deposition of tetracycline in growing bone and teeth, are a yellow-brown discoloration of the teeth and a reversible slowing of bone growth in children younger than 5 years of age. In addition, tetracyclines, especially demethylchlortetracycline, frequently cause photosensitivity with consequent erythema when the skin is exposed to sunshine.

A secondary effect, specific to minocycline, is the alteration of vestibular functions, with dizziness, tinnitus, and nausea.

5.4. Aminoglycosides (Aminocyclitols)

This is a large family of antibiotics produced by several strains of *Streptomyces*, *Micromonospora*, and *Bacillus*. Aminoglycosides are quite similar in their chemical characteristics, general biological properties, and mechanism of action.

In terms of chemical structure, they are aminocyclitols (cyclohexane with hydroxyl and amino or guanidino substituents) with glycosyl substituents at one or more hydroxyl groups. Because of this structure, the molecule is basic and has a high degree of solubility in water and poor solubility in lipids. These properties make it clear why the aminoglycoside antibiotics are poorly absorbed when given orally and suggest that their transport across the bacterial membrane must take place through a specific active transport mechanism and not through simple passive diffusion.

The aminoglycoside antibiotics can be classified according to the nature of the aminocyclitol, as shown in Figure 5.22.

5.4.1. Streptomycin

5.4.1.1. Properties

Streptomycin (Figure 5.23) was discovered in 1944 as the result of a research program carefully planned to isolate an antibiotic active against

Aminocyclitol		Main antibiotics
Name	Structure	
Streptidine	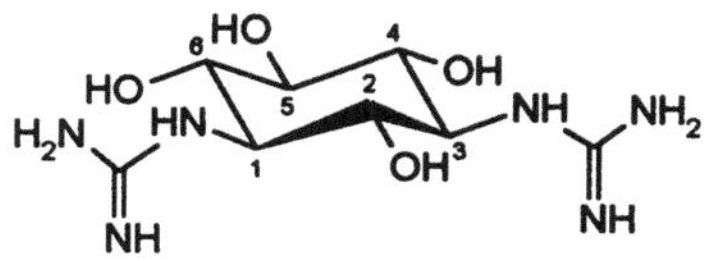	streptomycin
2-Deoxystreptamine	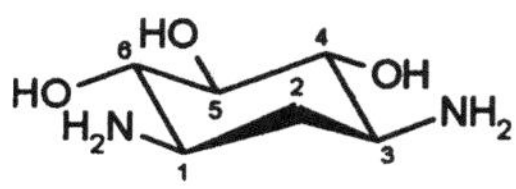	neomycin paromomycin ribostamycin kanamycin tobramycin gentamicin sisomicin amikacin dibekacin netilmicin
Actinamine	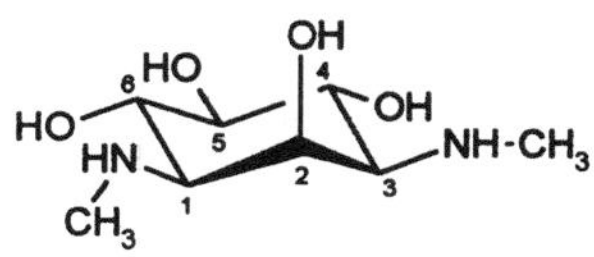	spectinomycin

Figure 5.22. Structures of aminocyclitol moieties of aminoglycoside antibiotics.

gram-negative bacteria, thus filling in the deficiencies of the spectrum of action of penicillin G. In addition to having excellent activity against these bacteria (except for some *Proteus* and some only slightly sensitive *Pseudomonas* species), streptomycin at low concentrations inhibits *Mycobacterium tuberculosis* and, consequently, was the first drug effective as a cure for tuberculosis. It is also active against staphylococci, whereas streptococci are rather insensitive.

The negative aspects that greatly limit its use today are

1. A specific toxicity for the auricular apparatus that results in vestibular damage with alteration of equilibrium and sometimes deafness
2. A rather high frequency of allergic reactions
3. A high frequency of resistant mutants and their widespread diffusion

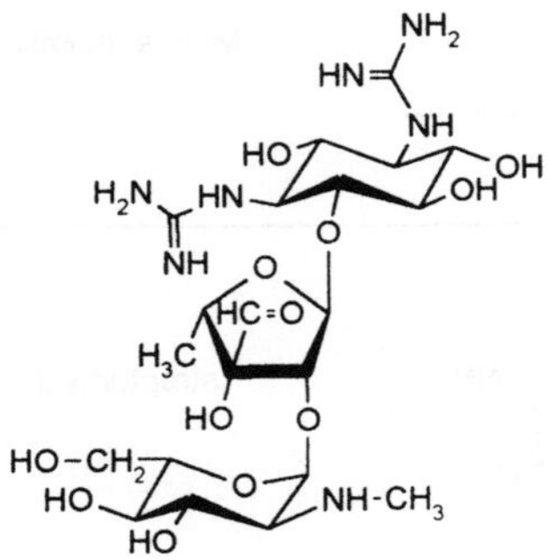

	MIC (μg/mL)
S.aureus	1
S.pneumoniae	0.2
E.faecalis	2
E.coli	1
P.mirabilis	4
P.aeruginosa	20
M.tuberculosis	0.5

Figure 5.23. Structure and *in vitro* activity of streptomycin.

There are different types of mutants resistant to streptomycin. The first type has an altered 30 S ribosomal subunit, which is the site of action of this antibiotic (see Section 3.5.3.2). These mutants are not resistant to other aminoglycoside antibiotics. The second type owes its resistance to decreased permeability to the antibiotic and these mutants tend to be cross-resistant to other aminoglycosides. Both types of mutants are present in populations of susceptible bacteria with different frequencies and can be selected during treatment of patients. However, the most widely found mechanism of resistance involves enzymatic inactivation, caused by transferable R factors. This reaction consists of adenylation or phosphorylation of the hydroxyl group at position 3 of methylglucosamine.

5.4.1.2. Structural Modifications

Elimination of the amidino groups, substitution of amine functions, and reductive amination of the aldehyde group all completely inactivate streptomycin. In contrast, demethylation of methylglucosamine, substitution of one of the amidino groups of streptidine with a carbamoyl (as in the antibiotic *bluensomycin*), and catalytic reduction of the aldehyde to a hydroxyl do not substantially change the activity. The last reaction produces *dihydrostreptomycin*, a compound that has also been isolated from a *Streptomyces* strain. It was introduced into therapy but later abandoned because of its ototoxicity. It is noteworthy that reduction of the aldehyde group decreases the frequency of hypersensitivity.

5.4.2. Aminoglycosides Containing 2-Deoxystreptamine

Many compounds are members of this group and several have clinical applications. 2-Deoxystreptamine itself has no biological activity. Substitution with various amino sugars in position 6 produces inactive compounds. When the substitution is in position 5 or 4, some antimicrobial activity emerges. Good activity appears only when positions 4 and 5 or 4 and 6 are both substituted.

There are thus two series of antibiotics containing 2-deoxystreptamine that are used in therapy. In the first, the aminocyclitol is substituted in two adjacent hydroxyl groups (positions 4 and 5); in the second, in two nonadjacent hydroxyl groups (positions 4 and 6) (Figure 5.24). In the first series, the sugar in position 5 is almost always ribose, and when an amino sugar is attached to it, the activity increases. In both series, the activity is influenced by the presence of amino groups on the sugar in position 4. There is maximal activity when there are two amines, in positions 2′ and 6′, and there is less activity when an amine is present only at one of these two positions. Another structural characteristic

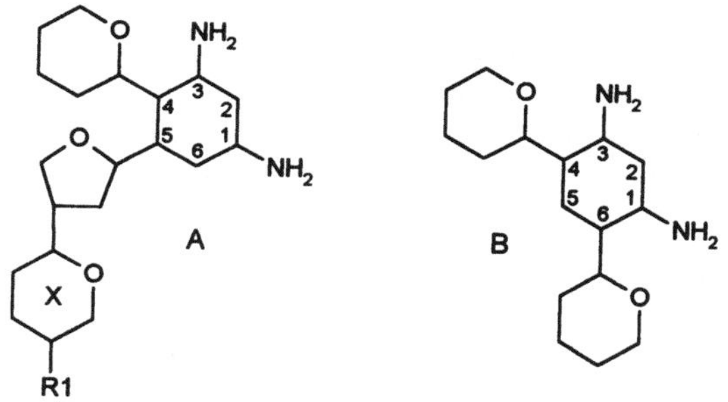

Figure 5.24. Schematic representation of the two classes of 2-deoxystreptamine-containing aminoglycosides. (A) 4,5-substituted. The X sugar is missing in the ribostamycins. (B) 4,6-substituted. In both classes the antibiotics differ in the amino, hydroxy, and methyl substituents on the sugars.

essential for activity is the presence of the two amino groups at positions 1 and 3 of 2-deoxystreptamine. Normally, any substitution of these has negative effects, but there are some specific cases, described later, in which a substitution on the amine at position 1 confers interesting properties to the molecule.

5.4.2.1. Aminoglycosides Containing 2-Deoxystreptamine Substituted in Positions 4 and 5

The best known products in this group are the *neomycins,* a mixture of antibiotics of which most preponderant is *neomycin B,* isolated in 1949 from *Streptomyces fradiae* (Figure 5.25). They have a spectrum of activity similar to that of streptomycin, including modest activity against *Enterococcus* and *Pseudomonas.* Neomycin is used only topically because it is ototoxic and nephrotoxic. *Paromomycin* (*aminosidine*) resembles neomycin in its chemical structure and biological properties.

Two antibiotics of this class have been developed in Japan for systemic use: *lividomycin A* and *ribostamycin* (Figure 5.25). The first has a good spectrum of activity, including *Pseudomonas,* and is less toxic than neomycin. The second is less active than the other deoxystreptamine derivatives (ribose is unsubstituted), but it is also less toxic.

5.4.2.2. Aminoglycosides Containing 2-Deoxystreptamine Substituted in Positions 4 and 6

This group of antibiotics (Figure 5.26) has become very important because of the high activity, both of the natural products and of some semisynthetic derivatives. The first compound of this group to be used in therapy was *kanamycin A,* which was isolated in 1957 from *Streptomyces kanamyceticus,* together with the *kanamycins B* and C. Its spectrum of activity is very similar to that of streptomycin: it is less toxic for the vestibular apparatus but affects hearing and is more nephrotoxic. It is used for gram-negative infections and in Japan is also used for treatment of tuberculosis. Kanamycin B, or *becanamycin,* has biological properties similar to those of kanamycin A and has also been introduced into clinical use.

In 1963 the *gentamicin* complex, which contains many components, was isolated from *Micromonospora purpurea.* The term gentamicin refers to the mixture commonly used in therapy, which contains *gentamicins C1, C1a,* and C2. They differ only in the degree of methylation of carbon 6′ of the sugar in position 4. Gentamicin is very active against *Staphylococcus aureus* and against gram-negative bacteria. Its excellent activity

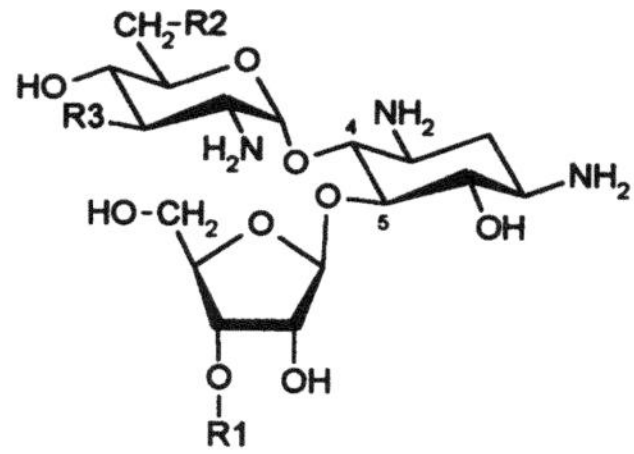

R1	R2	R3		MIC (μg/mL)			
				S.aureus	*E.coli*	*P.aeru-ginosa*	*Proteus* indole+
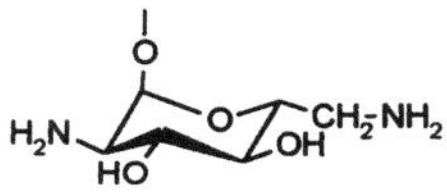	NH_2–	OH	Neomycin B	0.5	2-5	5-30	4
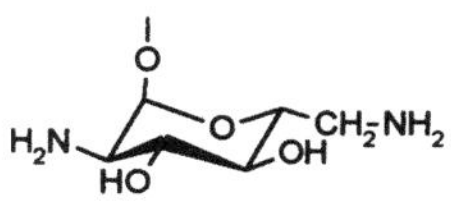	OH	OH	Paromomycin	1	8	>100	4
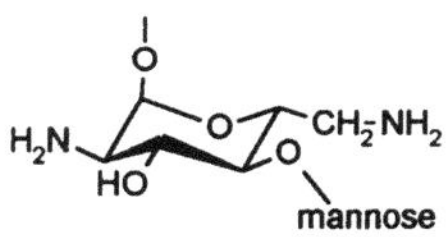	H	H	Lividomycin A	0.8	6	6	3
H	NH_2–	OH	Ribostamycin	12.5	3.6	>100	3-6

Figure 5.25. Aminoglycosides containing 2-deoxystreptamine substituted in positions 4 and 5.

against *Pseudomonas* and several *Proteus* species is of particular importance.

Compounds later introduced into clinical use are *tobramycin* and *sisomicin*. The first is produced by *Streptomyces tenebrarius* and has a structure similar to that of kanamycin B (it is 3′-deoxykanamycin B) but a spectrum of activity similar to that of gentamicin. It is more active than gentamicin against *Pseudomonas* and somewhat less active against the other gram-negative bacteria. Sisomicin is produced by *Micromonospora inyoensis* and has a structure and biological properties similar to those of gentamicin.

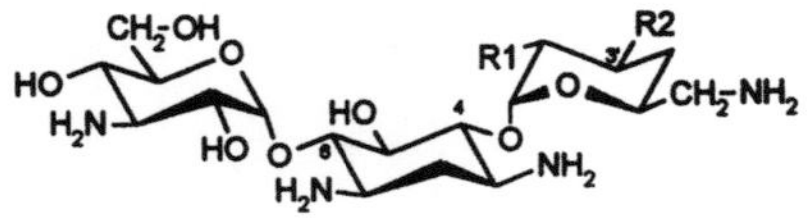

MIC (μg/mL)

R1	R2	Name	*S.aureus*	*E.faecalis*	*E.coli*	*Klebsiella sp.*	*P.aeruginosa*	*P.mirabilis*	*Proteus* indole+
OH	OH	Kanamycin A	0.5-2	8-32	3	0.2-2	>100	4	2-8
NH_2-	OH	Kanamycin B	0.8	-	0.8	-	50	-	1.5-3
NH_2-	H	Tobramycin	0.2	2-8	0.4	0.4	0.4	0.6	0.8

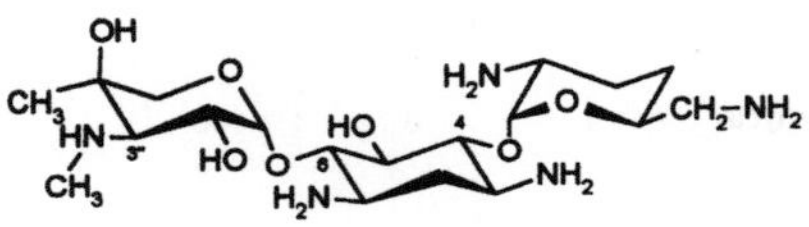

Gentamicin C1a

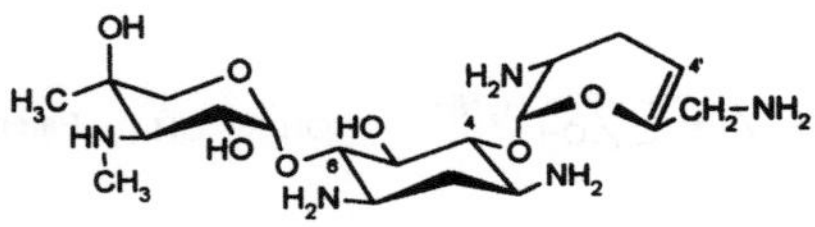

Sisomicin

MIC (μg/mL)

Name	*S.pyogenes*	*E.faecalis*	*E.coli*	*Klebsiella sp.*	*P.aeruginosa*	*P.mirabilis*	*Proteus* indole+
Gentamicin C1a	0.1	2.8	0.2	0.2	0.2-1.5	0.6	0.4-1.5
Sisomicin	-	-	0.4	0.2	0.2	0.6	0.6

Figure 5.26. Aminoglycosides containing 2-deoxystreptamine substituted in positions 4 and 6.

5.4.2.3. Relationships between Structure and Enzymatic Inactivation for the Semisynthetic Derivatives of Aminoglycosides Containing 2-Deoxystreptamine

To understand the relationship between the structures of the derivatives of 2-deoxystreptamine and their activity against certain bacterial strains, it is necessary to examine the mechanisms of resistance or of insusceptibility to these antibiotics.

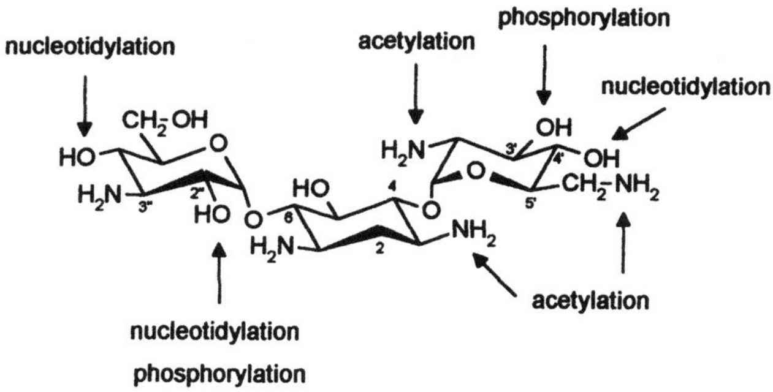

Figure 5.27. Representation of points of inactivation of kanamycin B by bacterial enzymes.

Unlike streptomycin, no cases of resistance related to alterations of the target have been found. Resistance related to inability to penetrate the bacterial cell is rarely observed, except in certain *Pseudomonas* strains, but quite frequently strains have been isolated that produce inactivating enzymes. A schematic representation is shown in Figure 5.27 of the modifications produced in kanamycin B by these enzymes.

Obviously, not all these modifications are of practical relevance. Acetylation of the nitrogen in position 6′ occurs rarely, adenylation of the oxygen in position 4′ and phosphorylation of the oxygen in position 2″ are carried out only by *S. aureus*. On the other hand, phosphorylation of the oxygen in position 3′ is quite common, especially by *Pseudomonas* strains. Two other important reactions are acetylation of the amine in position 3 of deoxystreptamine and nucleotide attachment to the oxygen in position 2″. One can therefore understand the activity of tobramycin (3′-deoxykanamycin B) against *Pseudomonas*, since this antibiotic does not possess the hydroxyl group in position 3′ and therefore cannot be inactivated by phosphorylation by the *Pseudomonas* enzymes. In the same way, gentamicin cannot be phosphorylated in position 3′ or have the nucleotide attached in position 4′ because neither of these two carbon atoms is hydroxylated.

This information served as the basis for the rational planning of synthesis of chemically modified derivatives. Among these, *amikacin* [1N-(α-hydroxy-γ-aminobutyryl)kanamycin A] and *dibekacin* (3′,4′-deoxykanamycin B) have been adopted for clinical use (Figure 5.28). Dibekacin is of course insensitive to the enzymes that attack the hydroxyl groups in positions 3′ and 4′ and therefore has activity comparable to that of gentamicin. The synthesis of amikacin was suggested by the observation that *butirosin B* (an antibiotic produced by *B. circulans* differing from ribostamycin only in having an α-hydroxy-γ-aminobutyric

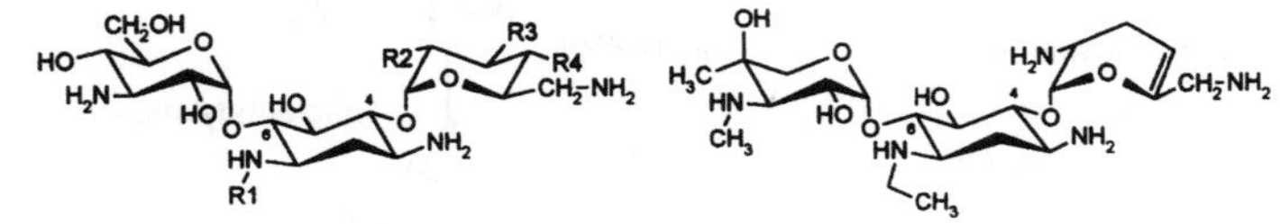

Amikacin and dibekacin — Netilmicin

R1	R2	R3	R4	Name	MIC (μg/mL) S.aureus	E.faecalis	E.coli	Klebsiella sp.	P.aeruginosa	P.mirabilis	Proteus indole+
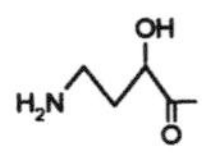	OH	OH	OH	Amikacin	0.4-3	2	2	2	3	3	1.5-3
H	NH_2	H	H	Dibekacin	0.8	3-6	1.5	-	-	-	3-12
				Netilmicin	0.1-0.8	0.2-12	0.2	0.2	0.2	1	0.2-12

Figure 5.28. Structures and activities of semisynthetic aminoglycosides.

acid substituent on the amine in position 1 of 2-deoxystreptamine) was active against *Pseudomonas* and other inactivating strains, unlike ribostamycin. The different aminoglycosides obtained by acylation with α-hydroxy-γ-aminobutyric acid were found to be insensitive to the enzymes phosphorylating the oxygen in position 3′, to those acylating the nitrogen in position 3, and to those determining nucleotide attachment to the oxygen in position 2″. Thus, amikacin, the derivative obtained by acylating the nitrogen on C-1 of kanamycin A, is the aminoglycoside with the broadest spectrum of activity. It is used especially in cases of infection caused by bacteria resistant to the other aminoglycosides.

A later semisynthetic derivative is *netilmicin* (1-*N*-ethylsisomicin) (Figure 5.28). Also in this case, substitution at the nitrogen protects the molecule from the attack of the inactivating enzymes and, consequently, netilmicin has a somewhat broader spectrum of activity than gentamicin and is active against gentamicin-resistant strains.

5.4.3. Aminoglycosides Containing Different Aminocyclitols

In addition to those already described, only one other aminoglycoside antibiotic is currently being used clinically. This is *spectinomycin* (Figure 5.29), which is produced by *Streptomyces spectabilis* and contains the aminocyclitol actinamine. It differs from the other derivatives in that it is bacteriostatic instead of bactericidal. It is used for the treatment of gonorrhea.

There are other aminoglycoside antibiotics that are used in agricul-

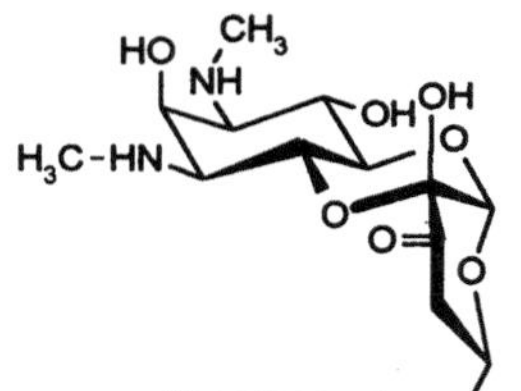

MIC (μg/mL)			
S.pyogenes	*S.pneumoniae*	*N.gonorrhoeae*	*E.coli*
6	12	8	12

Figure 5.29. Structure and *in vitro* activity of spectinomycin.

ture. Probably the most important is *kasugamycin*, used in rice cultivation in Japan for control of *Piricularia oryzae*. *Hygromycin B*, produced by *Streptomyces hygroscopicus*, and *destomycin*, produced by *Streptomyces rimofaciens*, are used to control ascarid infestation in animals,. Finally, *validomycin*, isolated from *S. hygroscopicus*, is used against some fungal diseases of plants.

5.4.4. Aminoglycosides Obtained by Biotransformation of Synthetic Precursors

By treating strains that produce aminoglycoside antibiotics with mutagens, it is possible to generate mutants that have lost their ability to synthesize 2-deoxystreptamine. These mutants are easily identified as they produce antibiotics only when 2-deoxystreptamine is added to the culture. Sometimes these microorganisms can utilize different exogenous aminocyclitols to produce the corresponding analogues of the natural antibiotic. This method for production of new antibiotics is called *mutational biosynthesis* or *mutasynthesis*.

The first products thus obtained were the *hybrimycins* (2-hydroxyneomycins), produced by mutant strains of *Streptomyces fradiae* grown in the presence of streptamine. Later, other microorganisms were used to synthesize new analogues of the antibiotics streptomycin, paromomycin, kanamycin, ribostamycin, butirosin, and sisomicin. The method had fruitful results. The products obtained, many of which have microbiological activity, have helped to clarify some structure–activity relationships and some of them may prove to be of interest for therapeutic use. The usefulness of this method is, however, limited by its low yields.

5.4.5. Unresolved Problems

The spectra of activity of the aminoglycosides were discussed earlier when describing the individual compounds and it was noted that, in general, this class of antibiotics lacks activity against anaerobic bacteria or aerobic bacteria growing under anaerobic conditions.

The problem of resistance has already been mentioned as well as the attempts to resolve it by planned synthesis of new products. There is no doubt that the new derivatives are an important therapeutic advance, and it is possible that current research efforts will yield further results.

The main limitation in the use of the aminoglycosides is represented by their nephrotoxicity and ototoxicity. Nephrotoxicity decreases from gentamicin to tobramycin, amikacin, and netilmicin, whereas the frequency of cochlear toxicity decreases from amikacin to gentamicin, tobramycin, and netilmicin. It has not yet been possible to elucidate the relationship between the structure of these antibiotics and the extent of their toxic effects.

5.5. Macrolides

Antibiotics with structures characterized by a large aliphatic lactone ring are called macrolides. They are conveniently divided into two classes with quite different biological properties and structures: the antibacterial macrolides and the antifungal macrolides also known as the polyene macrolides or *polyenes*. There are also macrolides possessing complex structures, e.g., the *avermectins*, used in agriculture and animal husbandry for their activity against nematodes and arthropods.

5.5.1. Antibacterial Macrolides

This group is characterized chemically by the comparatively smaller size of the lactone ring, from 12 to 16 atoms (Figure 5.30; macrolides with a 12-atom ring do not have practical applications), and biologically by their mechanism of action, i.e., specific inhibition of bacterial protein synthesis, resulting from formation of a complex with the ribosomal 50 S subunit.

The lactone ring has methyl and hydroxyl substituents in various

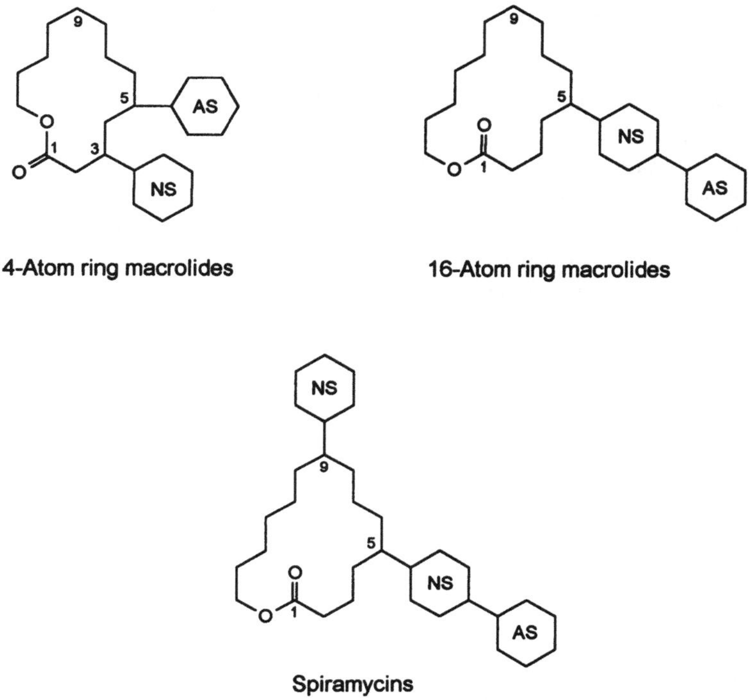

Figure 5.30. Schematic representation of the structures of antibacterial macrolides. AS = amino sugars; NS = neutral sugars; all other substituents and glycosidic bonds have been omitted.

positions. At least one glycoside substituent is always present, usually in position 5, and frequently a second one in position 3. The macrolides can be either neutral or basic, depending on whether the glycosides are ordinary sugars or amino sugars (the latter are more common).

The spectrum of antibacterial activity resembles that of penicillin G, with good activity against gram-positives and some gram-negative cocci, but, in addition, they are active against *Legionella*, mycoplasmas, and *Chlamydia*. With some exceptions, discussed later, they show cross-resistance among themselves and with lincomycin. Some of them have been introduced into clinical use in several countries. They are: *erythromycin*, *carbomycin* (no longer used), *oleandomycin*, *spiramycin*, the *kitasamycins* (or leucomycins), *josamycin*, *midecamycin*, and one derivative of the latter, *miocamycin*. In the veterinary field much used is *tylosin*, which is active against mycoplasmas.

5.5.1.1. Macrolides with a 14-Atom Ring

Erythromycin. This is the most widely used of the macrolide antibiotics. It was isolated in 1952 from fermentations of *Streptomyces erythreus* (now classified as *Saccharopolyspora erythrea*) as a complex of three very similar substances, *erythromycin A, B,* and *C,* of which the most active and major component is erythromycin A. Its structure (Figure 5.31) is typical of the macrolides with a 14-atom ring: it has no double bonds, has methyl groups on the even-numbered carbon atoms, and a carbonyl in position 9. Position 3 is substituted with cladinose and position 5 with desosamine.

Erythromycin is quite active against gram-positives (Figure 5.32) and is used for streptococcal and pneumococcal infections. It is less active against staphylococci, which are often resistant. It is active against some gram-negatives, such as *Neisseria, Haemophilus,* and *Bordetella,*

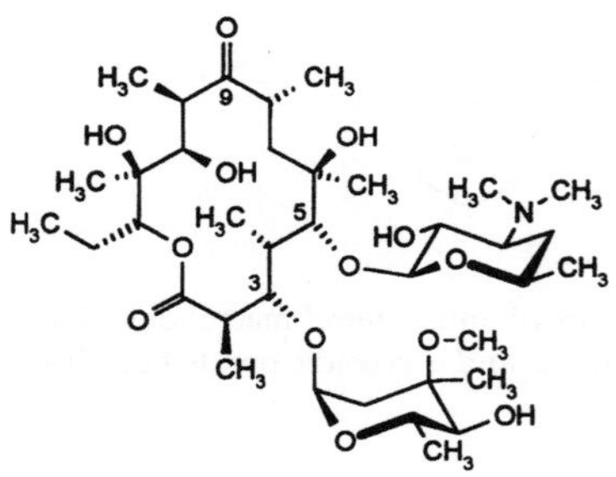

Erythromycin A

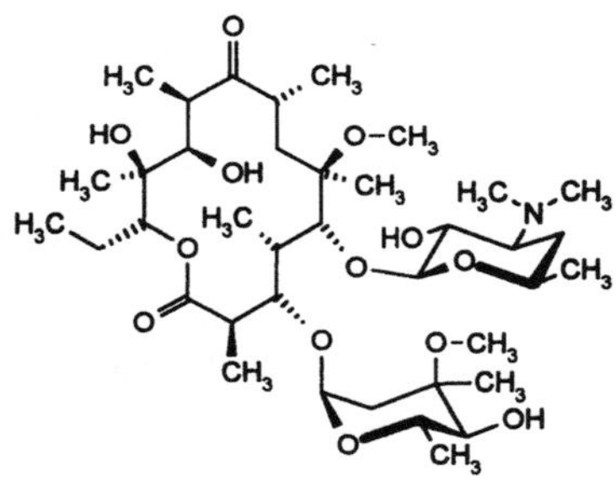

Clarithromycin

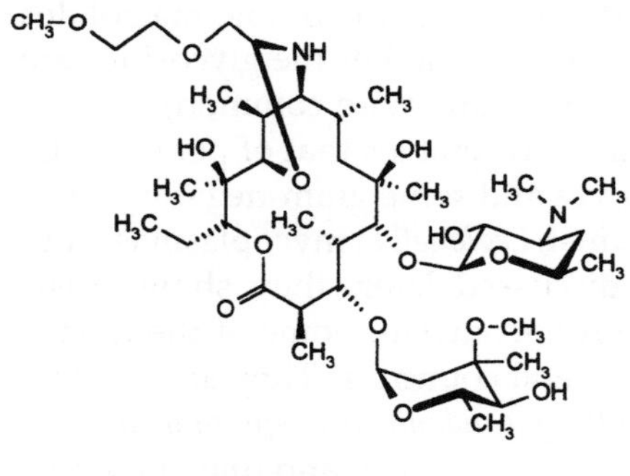

Dirithromycin

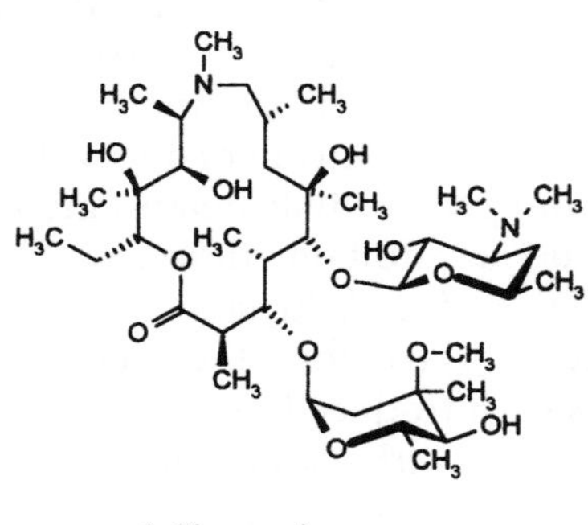

Azithromycin

Figure 5.31. Structures of macrolides with a 14-atom ring.

Name	MIC (μg/mL)				
	S.aureus	*S.pyo-genes*	*P.acnes*	*N.gonor-rhoeae*	*C.tracho-matidis*
Erythromycin	0.06	0.03	0.03	0.1	0.5
Oleandomycin	0.8	0.2	-	3	-
Flurithromycin	0.6	0.03	0.03	0.12	-
Clarithromycin	0.03	0.02	0.03	0.3	0.1
Dirithromycin	0.12	0.2	0.5	2	4
Azithromycin	0.06	0.03	0.004	0.03	0.3

Figure 5.32. *In vitro* activities of macrolides with a 14-atom ring.

whereas the activity against *E. coli*, demonstrable *in vitro*, is inadequate for therapeutic purposes.

Erythromycin is usually given orally but, because it is not sufficiently stable in an acidic environment (the first macrolide used in therapy, carbomycin, had to be abandoned because its absorption was too unreliable), blood levels are erratic. On the other hand, intramuscular administration of, e.g., erythromycin ethylsuccinate is painful. The laurylsulfate of erythromycin propionyl ester, called *erythromycin estolate*, gives blood levels at least three times higher than the base, but since the propionic ester is inactive, only the fraction hydrolyzed *in vivo*, which is difficult to measure quantitatively, can be taken into consideration. In addition, with the estolate, some cases of hepatotoxicity have occurred, almost certainly the result of hypersensitivity.

Oleandomycin. Isolated in 1954 from *Streptomyces antibioticus*, oleandomycin is a macrolide similar to erythromycin in structure and in biological properties, but on the whole somewhat less active. It is given orally in an acetylated form, triacetyloleandomycin, which is absorbed better than the natural product. It has a limited use, as it offers no advantages over erythromycin.

Structure–activity relationships and new derivatives. Several semisynthetic derivatives have been prepared and their activities compared with those of the natural erythromycins. It was found that:

1. Both glycoside substituents are needed for activity.

2. Modification of the desosamine, especially of the amino group, leads to inactive products, probably because this group is involved in the binding to the ribosomes (esters of the adjacent hydroxyl at 2′ were inactive *in vitro* but active *in vivo* as they are easily hydrolyzed).
3. Cladinose does not appear to be directly involved in the binding to the ribosomes, since it can be partly modified without inactivation.
4. Modification of the functions in the positions from 9 to 12 of the lactone ring, such as reduction or substitution of the carbonyl group, elimination of the hydroxyl group at 12, and acetylation in 11 give products with some activity.

As the inactivation in acidic medium, which is the main cause of erratic oral absorption of erythromycin, is related to the formation of a hemiketal ring between the carbonyl in position 9 and the hydroxyl in position 6, some derivatives have been synthesized, in which this reaction is hindered. Of some interest is *flurithromycin*, prepared by a combination of chemical modifications and biological transformations. The presence of a fluorine atom in position 8 deactivates the adjacent carbonyl.

Two recent derivatives are *clarithromycin* (6-deoxy-6-methoxy-erythromycin) and *dirithromycin*, in which a nitrogen atom substituting for the carbonyl oxygen in 9 forms a tetrahydrooxazine ring with the oxygen in position 11 (Figure 5.31). Clarlthromycin produces a lower frequency of gastrointestinal effects than erythromycin and can be administered twice a day instead of four times, because of more prolonged blood levels. In addition, it shows a better activity against *Legionella pneumophila*, mycoplasmas, and chlamydiae (Figure 5.32). Dirithromycin does not have a better antibacterial activity but has a longer half-life and a good tissue distribution, which may permit single daily administration.

A more substantial modification of the structure has been applied in *azithromycin*, in which the lactone ring has been expanded by the insertion of a nitrogen atom. This change confers to the antibiotic stronger basic properties and greater activity against some gram-negatives, such as *H. influenzae* and neisseriae, and against chlamydiae. In addition, it displays a better pharmacokinetics and a lower frequency of gastrointestinal effects.

5.5.1.2. Macrolides with a 16-Atom Ring

As briefly discussed in Chapter 4, a mechanism of resistance to macrolides consists of an alteration (methylation) of the ribosomal RNA.

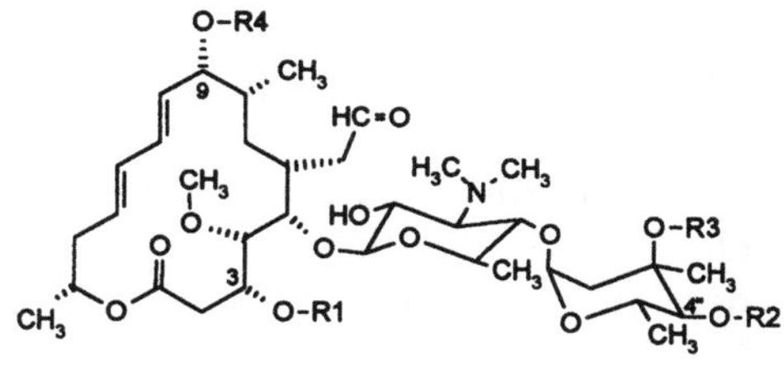

	R1	R2	R3	R4	MIC (µg/mL) *S.aureus*	*S.pyogenes*	*D.pneumoniae*	*E.faecalis*	*N.gonorrhoeae*
Leucomycin A1	H	isovaleryl	H	H	0.4	0.8	0.1	1.6	0.4
Josamycin	acetyl	isovaleryl	H	H	0.4	0.2	0.1	1.6	0.8
Midecamycin	propionyl	propionyl	H	H	0.4	0.4	0.05	1.6	1.6
Miocamycin	propionyl	propionyl	acetyl	acetyl	0.4	0.4	0.2	2	0.4
Spiramycin	H	H	H	forosaminyl	0.8	0.4	0.05	0.8	1.6

forosaminyl =

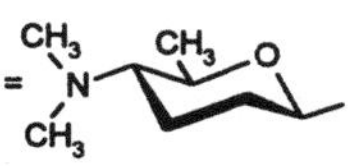

Figure 5.33. Structures and activities of macrolides with a 16-atom ring.

This change is induced by low concentrations of erythromycin or oleandomycin but not of other macrolides. Once induced, the bacteria are resistant to all macrolides and to lincomycin. When the inducer is removed from the culture, the strain reverts to susceptibility. This type of resistance has been found mainly in staphylococci. Macrolides with a 16-atom ring do not induce this type of resistance and they are active against several strains of *S. aureus* resistant to erythromycin. This fact has suggested their therapeutic use although they are less active than macrolides with a 14-atom ring (Figure 5.33).

Leucomycins. Most of the macrolides with a 16-atom ring used in therapy belong to the family of leucomycins, which is characterized by one pair of conjugated double bonds in the ring, an acetaldehyde group in position 6, and one disaccharide in position 5 (Figure 5.33). Leucomycins were initially isolated in 1953 from *Streptomyces kitasatoensis* as a mixture of several components. The main component is *leucomycin A1 (kitasamycin)*, which is in therapeutic use only in Japan.

More frequently used are other derivatives, *josamycin* and *mideca-*

mycin, produced by *S. narbonensis* and *S. mycarofaciens*, respectively, which differ from leucomycin A1 in the acyl substituents at the oxygen atoms in positions 3 and 4″. Josamycin is identical to leucomycin A3. These products show a variable oral absorption and do not give high blood levels, but they are well distributed in the tissues. The diacetyl derivative of midecamycin has been synthesized to improve the absorption and has been introduced in therapeutic use with the name *miocamycin*.

Spiramycin. This is produced by *Streptomyces ambofaciens* and differs from the leucomycins in having an additional amino sugar on the oxygen in position 9. Its activity is similar to but lower than that of the other macrolides. However, it is quite effective both against the experimental infections and in clinics, probably because of its good tissue distribution.

5.5.1.3. Unresolved Problems

A general characteristic of these macrolides is their low toxicity, so that they rarely show serious secondary effects. The most common is gastric intolerability. Cases of hepatitis have been seen with erythromycin estolate and very rarely with other macrolides. Allergic reactions are rare.

Oral absorption and pharmacokinetics of the natural derivatives are at present unsatisfactory but the more recent derivatives appear to overcome this problem. From a bacteriological point of view, the major problem seems to be the spread of resistant strains. In particular, macrolides are generally inactive against methicillin-resistant strains of *S. aureus* and frequently are also inactive against staphylococcal producers of penicillinase.

Interesting is the activity of the most recent derivatives against atypical mycobacteria, agents that frequently infect immunosuppressed patients.

5.5.2. Antifungal Macrolides (Polyenes)

These are structurally characterized by the large size of the lactone ring, from 25 to 37 carbon atoms, and by the presence of a series of conjugated double bonds, from 3 to 7, from which derives the name polyenes, in general, and trienes, tetraenes, etc., in particular. A general characteristic of this class is the presence in the ring of a series of hydroxyl groups, in positions opposite to the double bonds (see, e.g., the structure of *amphotericin B*, Figure 5.34), so that the molecule presents two distinct regions, one hydrophilic and the other hydrophobic. There-

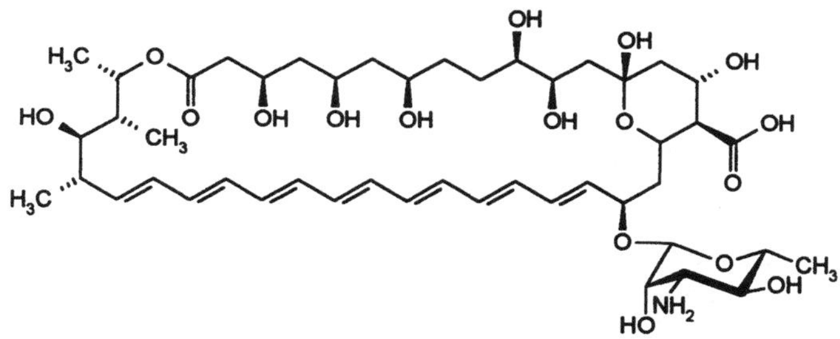

Figure 5.34. Structure of amphotericin B.

fore, these antibiotics display surfactant properties. Other common substituents are a carboxyl group and a sugar, mycosamine.

Polyenes are typically active against fungi, sometimes against protozoa, and only exceptionally against bacteria. This is a consequence of their mechanism of action, which, as discussed in Section 3.6.1.1, involves alteration of the integrity and function of the cell membrane through formation of complexes with the sterols that are components of eukaryotic membranes.

The antimicrobial activity increases with the number of conjugated double bonds. The available data are not sufficient to establish whether or not their toxicity, which is quite high, increases in parallel with their activity. It has been shown that there is no correspondence between hemolytic activity and antifungal activity. Several polyenes are used for local treatment of infections of the skin and the mucosa, and of intestinal infections since after oral administration they are not absorbed into the circulation. Among these are *pymaricin*, *tricomycin*, *candicidin*, and *nystatin*. The last is by far the most widely used. It is a tetraene, but with respect to both the size of the ring and other structural details it significantly resembles the heptaenes.

5.5.2.1. Amphotericin B

This is a heptaene with a 38-atom ring (Figure 5.34), produced by *Streptomyces nodosus*. It shows a better activity–toxicity ratio than the other polyenes, as a result of higher affinity toward ergosterol, the essential component of the yeast membrane, in comparison to cholesterol, present in the cell membrane of mammals. It is very active, with a cidal effect, against several pathogenic fungi, such as *Candida*, *Cryptococcus*, and *Histoplasma*, and less active against filamentous fungi, such as *Trichophyton*. It is administered intravenously to treat severe, life-threatening systemic mycoses. Its use is limited to severe infections because of its high toxicity, particularly renal toxicity.

5.5.2.2. Chemical Modifications

As the structural features of polyenes responsible for activity and toxicity are the same, not much has been done to improve their therapeutic index by chemical modifications. Esters of amphotericin and other macrolides have been synthesized; they retain their antifungal activity, but have not shown, in clinics, a better activity–toxicity ratio. Liposome-encapsulated amphotericin is reported to retain activity while showing lower toxicity.

5.5.2.3. Resistance

Although it is possible in the laboratory to obtain strains resistant to the polyene macrolides (with cross-resistance among them), from a clinical point of view resistance has not yet presented major problems. However, some strains resistant to *Candida* have been recently isolated from immunocompromised patients subjected to prolonged treatment. The resistant strains show an altered cell membrane structure.

5.6. Ansamycins

The ansamycins are characterized by a cyclic structure consisting of an aromatic moiety and an aliphatic chain (called the "ansa"). Unlike the macrolides, the ring is closed by an amide group, so that they are lactams rather than lactones.

They can be divided, on the basis of the aromatic moiety, into benzene ansamycins and naphthalene ansamycins (Figure 5.35). The benzene ansamycins, which include *geldanamycin* and *maytansine*, were isolated later than the naphthalene ones, from which they differ in their marked cytotoxicity. They have been studied not as antibacterial but as potential antitumor agents.

The naphthalene ansamycins can be divided into two groups according to the length of the ansa: a 23-atom or 17-atom ansa. Those with a 17-atom ansa are biologically characterized by their mechanism of action, i.e., specific inhibition of bacterial RNA polymerase. Depending on the producing microorganism and some structural details, these ansamycins are divided into the *streptovaricins* (in which the aliphatic chain is directly attached to the aromatic nucleus by a carbon–carbon linkage)

BENZENE ANSAMYCINS

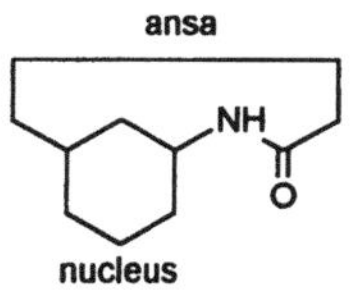

Geldanamycin
Maytansin

NAPHTHALENE ANSAMYCINS

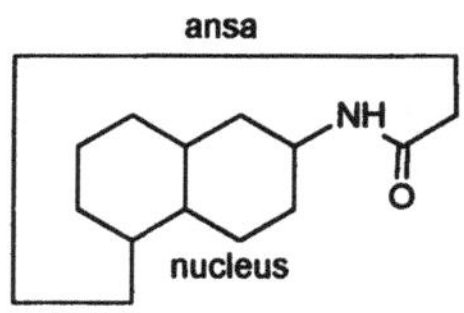

Streptovaricins
Rifamycin W

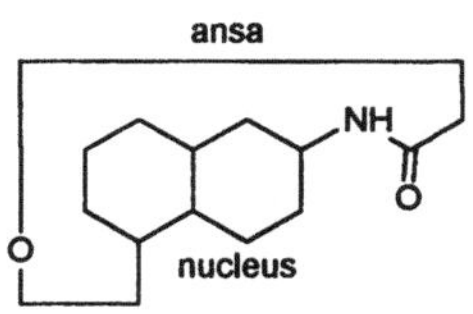

Rifamycins
Tolipomycins
Halomycins

Figure 5.35. Schematic representation of ansamycin structure.

and the *rifamycins, tolipomycins,* and *halomycins* (in which an ether bond interrupts the continuity between the aliphatic chain and the aromatic moiety).

The rifamycins constitute the most important family of the ansamycins. One semisynthetic rifamycin, *rifampin,* is the most active antibiotic available today for treatment of tuberculosis.

5.6.1. Natural and Semisynthetic Rifamycins

The first naturally occurring rifamycins were isolated in 1958, as a complex of five active substances, from fermentations of *Streptomyces mediterranei* (later classified as *Nocardia mediterranea* and more recently as *Amycolatopsis mediterranei*). Under specific culture conditions it is possible to obtain from the fermentation a single component, *rifamycin B* (Figure 5.36), which is moderately active against gram-positive bacteria and mycobacteria and shows low toxicity. The observation that in dilute

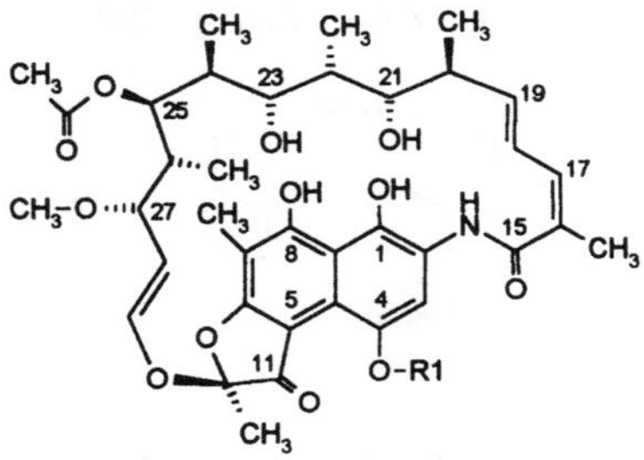

	R1	MIC (μg/mL)							
		S.aureus	*E.fae-calis*	*S.pneu-moniae*	*K.pneu-moniae*	*E.coli*	*P.aeru-ginosa*	*Proteus vulgaris*	*M.tuber-culosis*
Rifamycin B	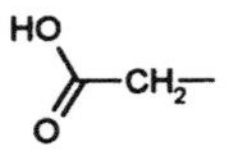	-	-	-	-	-	-	-	-
Rifamycin SV	H	0.005	0.05	0.025	25	50	50	50	0.25
Rifamide	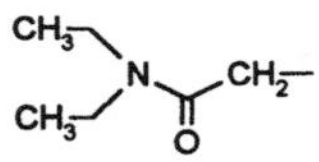	0.01	0.1	0.02	20	10	50	20	0.2

Figure 5.36. Structures and *in vitro* activities of rifamycin B, rifamycin SV, and rifamide.

solutions its activity increases with time, which suggests a transformation into more active derivatives, prompted a vast program of preparation of semisynthetic rifamycins. The first of these to be introduced into therapy was *rifamycin SV* (Figure 5.36), obtained by removing the glycolic chain in position 4 of rifamycin B by oxidative cyclization, hydrolysis, and reduction of the quinone thus obtained. Rifamycin SV can today be considered a natural rifamycin, since programmed research guided by biosynthetic considerations led to the isolation of a mutant of *Amycolatopsis mediterranei* that directly produces it. Rifamycin SV is very active and rapidly bactericidal against gram-positive bacteria and mycobacteria, less active against gram-negatives. It is well tolerated and is used parenterally, especially for infections of the bile tract where it reaches high enough concentrations to be effective even against the gram-negatives.

The limitations of rifamycin SV (which have dictated objectives for continuing research into semisynthetic rifamycins) are essentially as follows: (1) lack of oral activity; (2) weak activity against gram-negative bacteria; and (3) rapid elimination in the bile, with consequent low tissue levels. This last characteristic was of particular importance in tuberculosis, for which rifamycin SV was found to be poorly effective in spite of its excellent *in vitro* activity.

The modifications made in different positions on the molecule enabled the following structure–activity relationships to be established:

1. Alterations of the hydroxyl groups on positions 21 and 23 of the aliphatic chain and 1 and 8 of the naphthalene nucleus always give inactive products (except for oxidation to carbonyl of the hydroxyl group at position 1).
2. Inactive products are also obtained from any structural modification that changes the spatial orientation of these hydroxyls, suggesting their participation in the formation of the bonds with RNA polymerase (see mechanism of action in Chapter 3).
3. Minor modifications of the chain, such as hydrogenation of the conjugated double bonds, deacetylation, etc., do not substantially modify the activity.
4. Substitutions on positions 3 and 4 of the aromatic nucleus do not alter the intrinsic activity of the molecule, but can modify the physicochemical properties, and consequently such biological properties as absorption, distribution in the body, and penetration into bacteria, especially the gram-negative ones. If the substituent includes a carboxyl group, the antibacterial activity is invariably low, whereas basic functions increase the activity against gram-negative bacteria.
5. Activity after oral administration is often found in derivatives with substitutions in position 3 or with cyclic substituents containing a nitrogen function in positions 3 and 4. It is occasionally found in those substituted only in position 4.

Practical results of this research include *rifamide* (Figure 5.36), which, although possessing the same limitations as rifamycin SV, has a better therapeutic index, and *rifampin* (called *rifampicin* outside the United States).

5.6.1.1. Rifampin

Rifampin (Figure 5.37) has been found to possess a high degree of the desired characteristics. It has a broad spectrum of activity, is absorbed when given orally, and, most important, because of its better distribution in the body, is also very effective for treatment of tuberculosis. Its efficacy is partly related to its ability to also inhibit bacteria

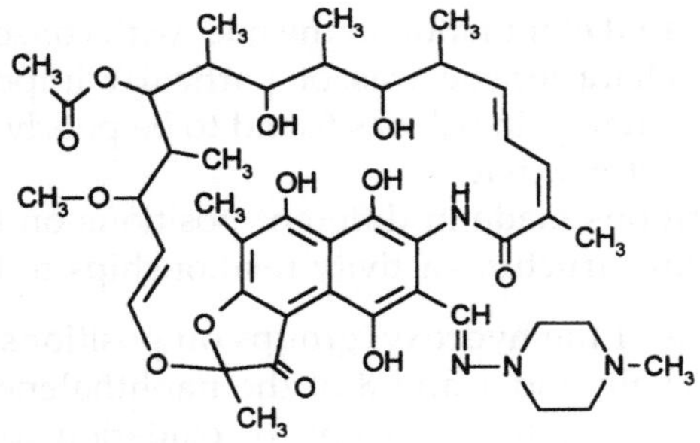

MIC (μg/mL)

S.aureus	E.faecalis	S.pneumoniae	K.pneumoniae	E.coli	P.aeruginosa	Proteus vulgaris	M.tuberculosis
0.002	0.01	0.01	5	1	10	5	0.05

Figure 5.37. Structure and *in vitro* activity of rifampin.

within cells such as macrophages or leukocytes. It is used throughout the world for treatment of tuberculosis and of leprosy.

Because of rifampin's high activity against staphylococci including methicillin-resistant ones, it is used in the severe infections caused by these bacteria, usually in combination with other antibiotics to prevent development of resistant mutants. It is also used for the prevention of meningitis caused either by *Neisseria meningitidis* or *Haemophilus influenzae*.

5.6.1.2. Other Rifamycins

Among the many semisynthetic rifamycins, some substituted in position 3 with long lipophilic chains have been found to have a certain degree of inhibitory activity against an enzyme typical of oncogenic viruses, the so-called reverse transcriptase (an enzyme used by RNA viruses to synthesize complementary DNA strains). Although many derivatives with analogous characteristics have been synthesized and their biological properties thoroughly studied, none has been developed clinically. There has also been no practical consequence of the observation that at high concentrations rifampin inhibits the reproduction of certain DNA viruses such as vaccinia.

Some derivatives, such as *rifabutine* (Figure 5.38), have shown activ-

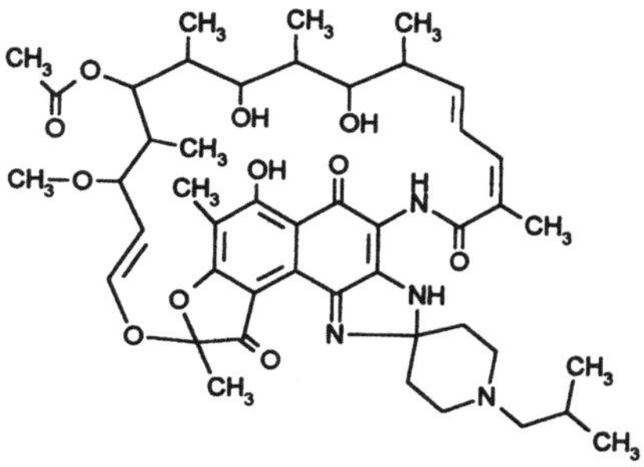

Rifabutine

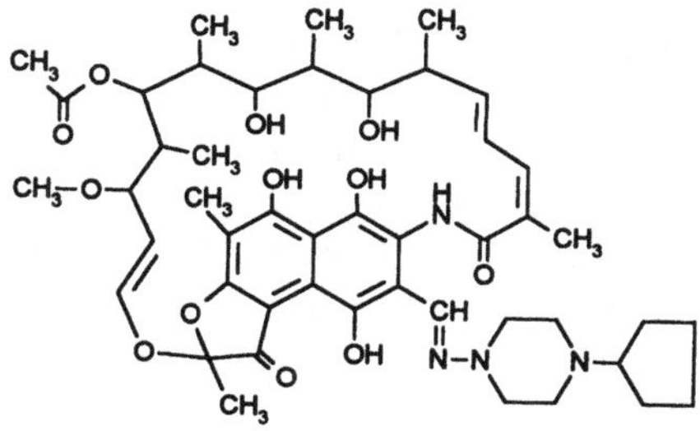

Rifapentine

Figure 5.38. Structures of two recent semisynthetic rifamycins.

ity against a fraction of the mutant strains resistant to rifampin. Another characteristic of rifabutine is its higher activity against atypical mycobacteria, such as *Mycobacterium avium*, so that it is used for the prevention of infection caused by these bacteria in immunosuppressed patients.

Another derivative, *rifapentine* (Figure 5.38), is under study for its pharmacokinetics, which suggests the possibility of administration at several-day intervals, and for its activity against atypical mycobacteria.

5.6.2. Unresolved Problems

5.6.2.1. Side Effects and Hypersensitivity

Rifampin, administered daily, rarely causes side effects except for an increase in bilirubin blood levels and a transient increase of hepatic transaminases, which are both considered harmless. However, in some patients treated over periods of many months, especially when treatment is intermittent, hypersensitivity reactions are observed; among them is the so-called flu syndrome, with fever, weariness, and joint ache. Very rare are severe cases of immunological reactions, which may include hemolytic anemia, renal block, and shock.

5.6.2.2. Resistance

There is generally a rather high frequency of mutants resistant to rifamycins in bacterial populations, of the order 10^{-7} in staphylococci and *E. coli*; in contrast, the frequency is much lower in *M. tuberculosis*.

The resistant mutants studied have usually been found to have alterations in RNA polymerase. Resistance associated with a transferable factor has never been demonstrated. This is probably why there has not been a marked increase in the spread of resistant strains after several years of therapeutic use.

The rifamycins show cross-resistance within the group. Some derivatives, as discussed previously, have been demonstrated to have some degree of activity against strains resistant to rifampin. It is unlikely that the problem of resistance to rifampin could be overcome through chemical modification. Today, it is believed that the problem may be circumvented by administering rifampin in suitable combinations with other antibiotics.

5.7. Peptide Antibiotics

As indicated by their name, these are compounds composed of amino acids connected through peptide bonds. They are of importance historically because almost all the early antibiotics studied for therapeutic use belonged to this class. Chemically they are a quite heterogeneous group and differ from proteins in being smaller, composed of at most about 30 amino acids, and in the following frequently found characteristics:

1. They may contain D-amino acids.
2. They may contain some uncommon, nonprotein, amino acids (e.g., *N*-methylamino acids, β-amino acids).
3. They often are cyclic molecules.
4. They contain heterocyclic rings, frequently thiazoles.

The group appears even more heterogeneous in its biological properties, including antimicrobial activity, mechanism of action, and toxicity.

The structure–activity relationships have been studied in detail for only a few families of peptide antibiotics. These are:

1. The penicillins and the cephalosporins, which have been discussed separately, but which can be formally considered as peptide antibiotics.
2. The actinomycins, which have a limited use as antitumor agents, and which are discussed in the section on antitumor antibiotics (see Section 5.9).
3. Glycopeptides or dalbaheptides, which, possessing particular characteristics, are described separately in Section 5.8.

5.7.1. Systemically Used Peptide Antibiotics

Most of these antibiotics are too toxic for systemic use. The following exceptions are noteworthy.

5.7.1.1. Polymyxin and Colistin

These are interesting examples of antibiotics active only against gram-negative bacteria. They are produced by various strains of bacilli and are very similar in their chemical structures (Figure 5.39) and biological properties. *Polymyxin B* is generally used as the sulfate for intramuscular or intravenous injections whereas the preferred form for colistin is its methane sulfonate derivative, which is less toxic but correspondingly less active. Both are used only exceptionally for treating severe infections by organisms such as *Pseudomonas* that are resistant to other less toxic antibiotics. The emergence of resistant strains is rare.

5.7.1.2. Capreomycin

This substance, isolated in 1960 from *Streptomyces capreolus*, is active against mycobacteria. Its MIC against *Mycobacterium tuberculosis* is somewhat high (about 10 μg/ml) but, since high blood concentrations are easily attained, the product can be used in the treatment of tuberculosis patients when the infecting strain is resistant to other less toxic agents.

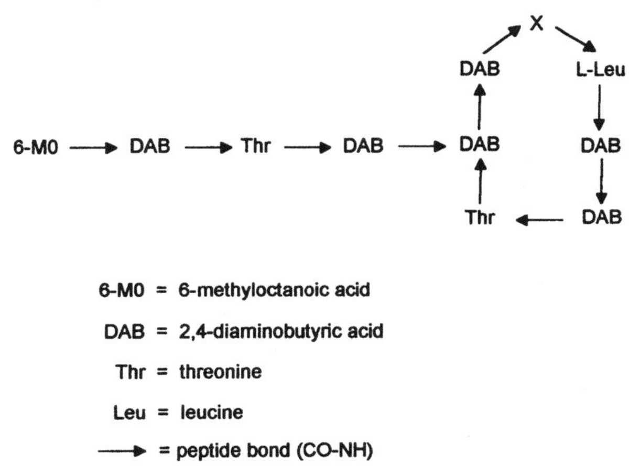

Figure 5.39. Structure of polymyxin B (X = D-phenylalanine) and colistin (X = D-leucine).

5.7.2. Topically Used Peptide Antibiotics

These include some of the earliest antibiotics studied for therapeutic use. *Tyrothricin*, isolated in 1939, is a mixture of two products, *gramicidin* (about 20%) and *tyrocidin* (about 80%). Gramicidin is very active against gram-positive bacteria, tyrocidin is less active but has some activity against gram-negatives also. Both are hemolytic agents, which precludes their systemic use. Topically, gramicidin is used for infections of the eye, nose, and throat and for ulcers and wounds.

Bacitracin, isolated in 1943, is active against gram-positive bacteria by its interference with cell wall synthesis. It can be given intramuscularly in exceptional cases but is normally considered too toxic (kidney damage) for systemic use. It is widely used as an additive to animal feed.

The search for new peptide antibiotics for clinical use is still very active as there are good reasons to believe that new effective members of the class could be found. In fact, two lipopeptide antibiotics are now under study, *daptomycin* and *ramoplanin*, for their marked activity against gram-positive bacteria. Furthermore, a new peptide antibiotic, *GE2270 A*, is promising also for its uncommon mechanism of action, the inhibition of the bacterial protein EF-Tu.

Several peptide antibiotics have been used outside clinical medicine, such as *nisin*, a food preservative, *siomycin*, and *thiostrepton*, used in animal husbandry.

5.8 Glycopeptide Antibiotics (Dalbaheptides)

This family of antibiotics is defined by its chemical structure and mechanism of action. They are linear heptapeptides, in which at least five of the amino acid residues are benzene rings connected to form one triphenylether and one diphenyl moiety. The aromatic rings bear various substituents, such as hydroxyl and methyl groups, chlorine atoms, and sugars. From a biological point of view, the striking aspect is their mechanism of action. They complex with the terminal D-alanyl-D-alanine of one or more intermediates of peptidoglycan synthesis, thus interrupting this process. The proposed name of *dalbaheptides* (D-alanyl-D-alanine-binding heptapeptide) is a combination of their chemical and biological characteristics.

Among the various antibiotics of this group, *vancomycin* and *teicoplanin* (Figure 5.40) are now used in clinics. In the past *ristocetin* was used, but then abandoned because of untoward hematic effects. *Avoparcin* is used as a feed additive for animals.

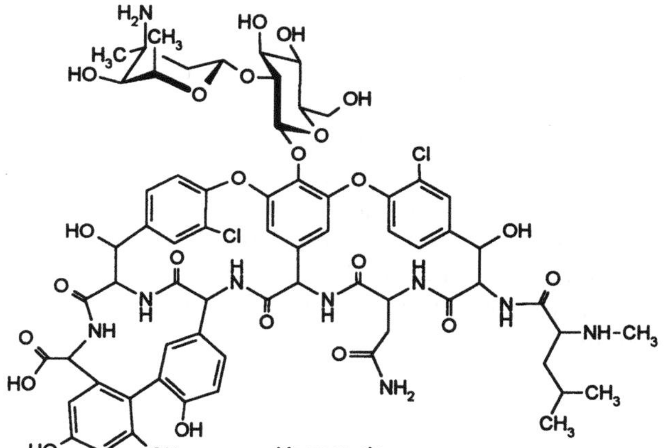

Vancomycin

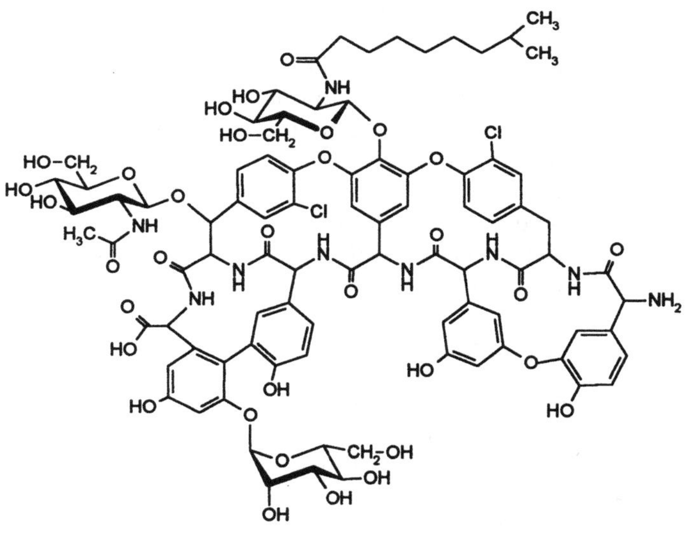

Teichomycin A2-2

	MIC (μg/mL)			
	S.aureus	*S.pyogenes*	*E.faecalis*	*C.perfringens*
Vancomycin	0.25	0.13	0.5	0.13
Teichomycin A2-2	0.13	0.06	0.13	0.003

Figure 5.40. Structures and activities of vancomycin and teicoplanin (teichomycin A2-2).

5.8.1. Vancomycin

Vancomycin was isolated in 1956 from fermentation of *Streptomyces orientalis* (now classified as *Amycolatopsis orientalis*). It is active against gram-positive bacteria on which it is normally bactericidal (Figure 5.40). Its use was limited for a long time, but it has become widespread in recent years because it is active against strains resistant to other antibiotics, in particular against methicillin-resistant staphylococci. According to its pharmacokinetics, it is administered daily in multiple doses by slow intravenous infusion, as rapid injections cause a dangerous histamine release (red man syndrome).

5.8.2. Teicoplanin

This antibiotic is produced by *Actinoplanes teichomyceticus* as a mixture of five components similar in their structures and biological properties, designated as *teichomycin A2–1* to *A2–5*. The most abundant component is teichomycin A2–2. The main characteristic of teicoplanin is the presence of an acyl chain attached to one of the sugars, which confers better pharmacokinetic properties to the molecule, so that it can be administered in daily single doses intravenously or intramuscularly. Teicoplanin does not cause histamine release. It is bactericidal and shows the same spectrum of action as vancomycin, but generally it is more active against *S. aureus* and streptococci and less active against coagulase-negative staphylococci (Figure 5.40).

5.8.3. Unresolved Problems

Both vancomycin and teicoplanin are generally well tolerated and do not show marked side effects. The early preparations of vancomycin were considered toxic and allergenic but it was later ascertained that this was the result of impurities. However, the tendency to produce the red man syndrome has been observed also with a pure product. Analogously, the recent preparations of vancomycin do not seem nephrotoxic, but it has been verified that vancomycin can worsen the nephrotoxic effect of aminoglycosides. These risks appear lower with teicoplanin.

Resistance is not very common. However, some enterococcal strains have been isolated, which show a moderate to high level of transferable resistance. The two antibiotics show an incomplete cross-resistance.

5.9. Antitumor Antibiotics

The antibiotics used for antitumor therapy belong to various chemical classes. Although they have different mechanisms of action, all induce irreversible damage to the DNA filaments. They are mostly cytocidal and act on growing cells only. The main compounds used in clinics are briefly described in the following subsections.

5.9.1. Actinomycins

The early members of this family of antibiotics were isolated in 1940 from *Streptomyces antibioticus*. Later, some tens of derivatives, which differ in the amino acids constituting the two side chains, were obtained from various microorganisms. Among these, *actinomycin D*, or *dactinomycin* (Figure 5.41), produced by *Streptomyces parvulus*, was used for

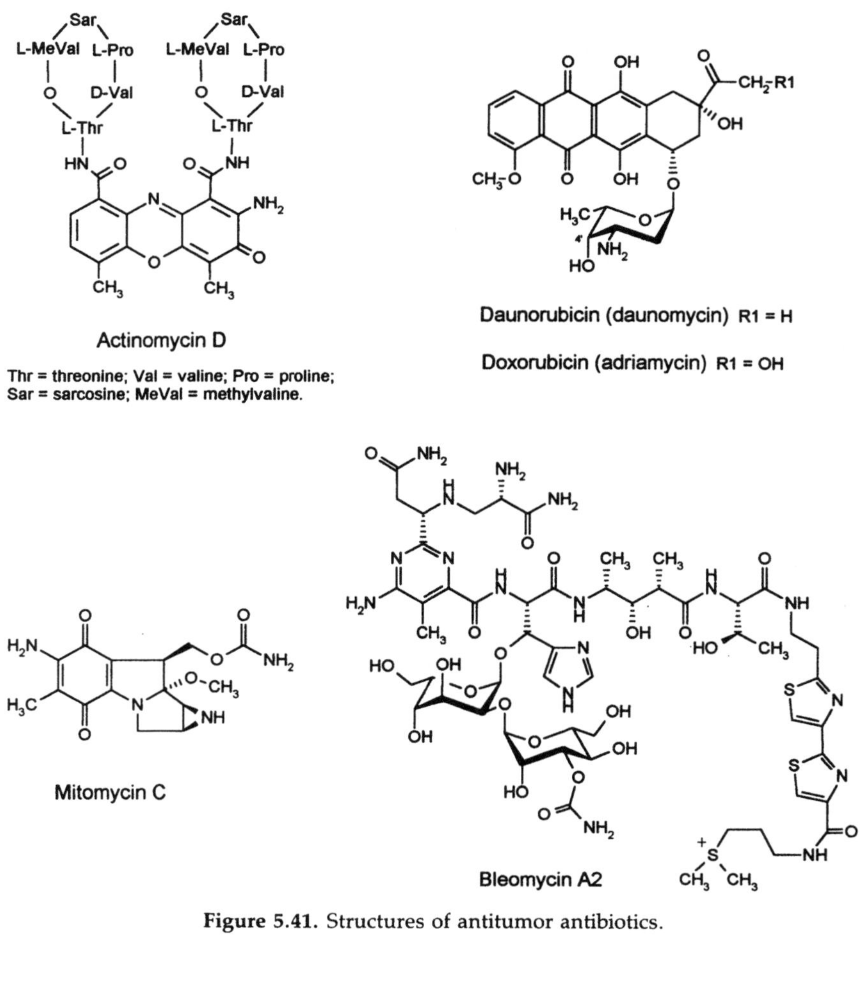

Figure 5.41. Structures of antitumor antibiotics.

the treatment of choriocarcinomas and sarcomas. Its importance is mostly historical as its use was limited by a marked toxicity for kidney, liver, and bone marrow.

5.9.2. Anthracyclines

The anthracyclines are the most effective compounds among the antitumor antibiotics. The main products of this family are *daunorubicin* and *doxorubicin*, also called *daunomycin* and *adriamycin*, respectively (Figure 5.41). The former is active mainly on leukemias and lymphomas. The latter has the broadest spectrum of action of all antitumor drugs and is effective in the treatment of a variety of tumors.

The main problem in their use is, in addition to the myelosuppression, the cardiac toxicity, which is inevitably observed when the cumulative dose exceeds a critical threshold. Resistant cells have been described that express in considerable amounts a membrane glycoprotein, called p170, responsible for an active export of the anthracyclines out of the cell.

The search for new derivatives able to overcome, at least in part, these drawbacks is very active. Among the new derivatives, *epirubicin*, an epimer of doxorubicin, which has the hydroxyl group at position 4′ in equatorial orientation, shows the same spectrum of action of the natural product but seems less toxic, probably because it is metabolized more rapidly. *Idarubicin*, a derivative of daunorubicin characterized by the absence of the methoxyl group on carbon 4 of the chromophore, is very lipophilic and thus orally active with the same indications as daunomycin.

5.9.3. Mitomycin C

Mitomycin C (Figure 5.41), isolated in 1958, is characterized by an aziridine cycle, which *in vivo* is activated by reduction and reacts with the guanidine bases to form covalent bonds with the opposite strands of DNA. It is mainly used in combination with other drugs in the therapy of solid tumors. Its main toxicity is myelosuppression. Resistant cells have been noted that show alterations in the repair mechanisms of DNA.

5.9.4. Bleomycin

Bleomycin is produced by *Streptomyces verticillus* as a mixture of nine components, which have been used as starting materials for the preparation of hundreds of semisynthetic derivatives. They are basic glycopeptides (Figure 5.41) that cause breaks in the DNA strands and are used in the treatment of lymphomas and tumors of testis and skin. The main toxicity is pulmonary fibrosis.

More than 100 analogues of bleomycin have been synthesized, and detailed studies on structure–activity relationship have been carried out. One of these derivatives, *peplomycin*, is active on various solid tumors and seems less toxic.

5.10. Miscellaneous Antibiotics

Grouped under this heading are some clinically used antibiotics with specific chemical or biological characteristics that do not fit into any of the classes described previously.

5.10.1. Chloramphenicol

Isolated in 1947 from a strain of *Streptomyces venezuelae*, chloramphenicol was the first therapeutic agent active against rickettsiae and related microorganisms, in addition to gram-negative and some gram-positive bacteria. It is well absorbed after oral administration and is the drug of choice for treatment of typhus and salmonellosis. The antibacterial spectrum of chloramphenicol is shown in Table 2.1.

Its widespread therapeutic use has been partly limited by the risk of bone marrow aplasia, a rare hypersensitive reaction (about one case in 40,000 treated) that is nearly always fatal. The occurrence of aplasia is not correlated with the dose but appears after long treatments and can be prevented by constant monitoring of the hematological condition of the patient. Less serious side effects include anemia, which is reversible and dose dependent, some gastrointestinal disturbances, and, because of the broad spectrum of the antibiotic's action, secondary infections resulting from elimination of the normal bacterial flora.

The structural formula of chloramphenicol is shown in Figure 5.42. As there are two asymmetric centers, four stereoisomers are possible, but only the D-*threo* isomer is active. Chloramphenicol is among the very few antibiotics produced industrially by synthesis rather than by fermentation.

From the numerous analogues that have been synthesized, the following structure–activity relationships could be deduced:

1. The dichloroacetic residue can be replaced by other acids without substantial loss of activity as long as they are strongly electronegative and do not cause higher steric hindrance.

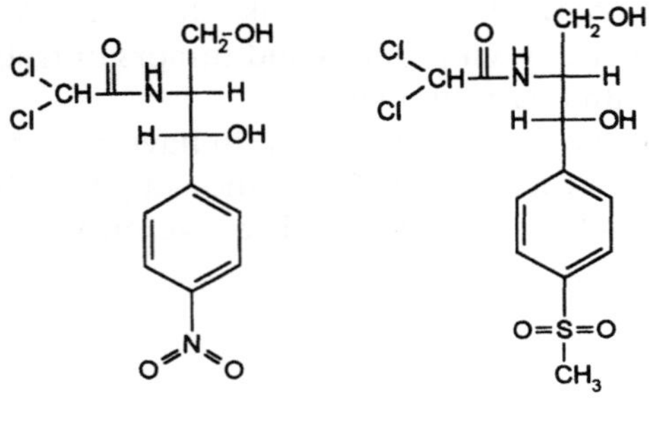

Figure 5.42. Structures of chloramphenicol and thiamphenicol.

2. The propanediol structure is essential for activity as indicated by the fact that only one stereoisomer has activity. Esters at the primary hydroxyl retain activity, probably because they are hydrolyzed *in vivo*. The chloramphenicol palmitate is used because, in contrast with the parent compound, it has no bitter taste.
3. In the aromatic moiety, the nitro group can be replaced by other polar functions, such as methylthio or methylsulfonic groups. The derivative obtained by this last replacement, *thiamphenicol* (Figure 5.42), has been introduced in therapy.

Resistance to chloramphenicol is usually related to inactivating enzymes that acylate one or both hydroxyls. These enzymes are inducible in *Staphylococcus aureus* and constitutive in *Escherichia coli*. In this latter, the enzyme is coded for by an R factor, which frequently carries the determinants for resistance to other antibiotics.

5.10.2. Lincomycin and Clindamycin

Lincomycin, isolated in 1962 from cultures of *Streptomyces lincolnensis*, is mainly active against gram-positive bacteria. Chemically, it can be considered an amino sugar acylated with a proline derivative (Figure 5.43). Its biological activity resembles that of the antibacterial macrolides (it inhibits protein synthesis by complexing with the ribosomal 50 S subunit) with which it has some degree of cross-resistance. It has been suggested that in spite of the apparent structural differences, the three-dimensional model of lincomycin has points of similarity with erythromycin. Three of the methyl groups of erythromycin (that on the ethyl,

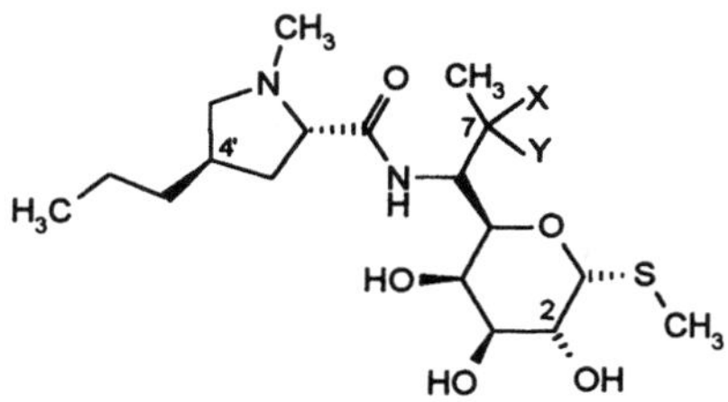

X	Y	Name	MIC (μg/mL) S.aureus	S.pyogenes	S.pneumoniae	E.coli
OH	H	Lincomycin	0.5	0.05	0.5	>100
H	Cl	Clindamycin	0.1	0.02	0.05	>100

Figure 5.43. Structures and activities of lincomycin and clindamycin.

that on carbon 2, and that in position 5 of the cladinose) may correspond spatially with three methyl groups of lincomycin (the S-methyl, the methyl in position 8, and the methyl on the propyl group). In addition, the hydroxyl in position 12 of the macrolides may coincide with the hydroxyl in position 2 of lincomycin.

Because of the simplicity of the molecule, several analogues have been prepared semisynthetically, with the following results:

1. The acyl portion can be modified without decrease of activity. In particular, the activity increases when the size of the alkyl group in position 4 of the proline is increased.
2. When the methyl on the sulfur is replaced by other alkyl groups, one obtains products that are more active *in vitro* but not *in vivo*.
3. Replacement of the hydroxyl in position 7 with a chlorine atom causes a marked increase in activity if the replacement takes place with inversion of the configuration. If the configuration is maintained, the activity is unchanged.

Clindamycin (Figure 5.43), prepared by means of this latter modification, has entered into therapeutic use because of its greater activity against gram-positive and anaerobic bacteria. It also has an interesting activity against *Plasmodium berghei*. The increased antibacterial activity of clindamycin seems to be related to its better penetration into the bacterial cell.

The lack of activity of these antibiotics against gram-negative organisms appears to be the result of a lower sensitivity of the ribosomes of gram-negative bacteria. The therapeutic limitations of these antibiotics are more or less the same as those for erythromycin. However, intramuscular administration of lincomycin is better tolerated. The most important side effect is pseudomembranous colitis, which sometimes can be very severe.

5.10.3. Novobiocin and Coumermycins

Novobiocin (Figure 5.44), discovered in 1956, is active against gram-positive bacteria and to a lesser extent against *Proteus* and other gram-negatives.

The *coumermycins*, chemically and biologically related to novobiocin, are a family of antibiotics discovered in 1965, which was subjected to a considerable program of chemical modifications. The most effective natural compound, coumermycin A, has also been studied in the clinic, but proved to be difficult to administer and was poorly effective.

5.10.4. Fusidic Acid

This antibiotic, active against gram-positive bacteria, was isolated from cultures of a fungus, *Fusidium coccineum*. It has a steroid-type structure (Figure 5.44), as do some other antibiotics produced by fungi (e.g., *cephalosporin P* and *helvolic acid*). It is absorbed orally and is mainly used in combination with penicillin. The several derivatives prepared have not revealed better activity than the natural product.

5.10.5. Fosfomycin

Initially isolated from a strain of *Streptomyces fradiae*, fosfomycin (earlier called phosphonomycin; Figure 5.44) is now produced synthetically, because of the simplicity of its structure. It is bactericidal and has a quite broad spectrum of action, which includes gram-positive bacteria, gram-negative enterobacteria, and some anaerobes, but its activity is generally moderate.

It is fairly well absorbed orally and can also be administered parenterally. The main limitation for its use in clinics is the high frequency of resistant mutants and it is therefore mainly used in combination with other antibiotics. Its toxicity is very low and secondary effects are rare, except for some adverse gastric effects and irritation at the site of injection.

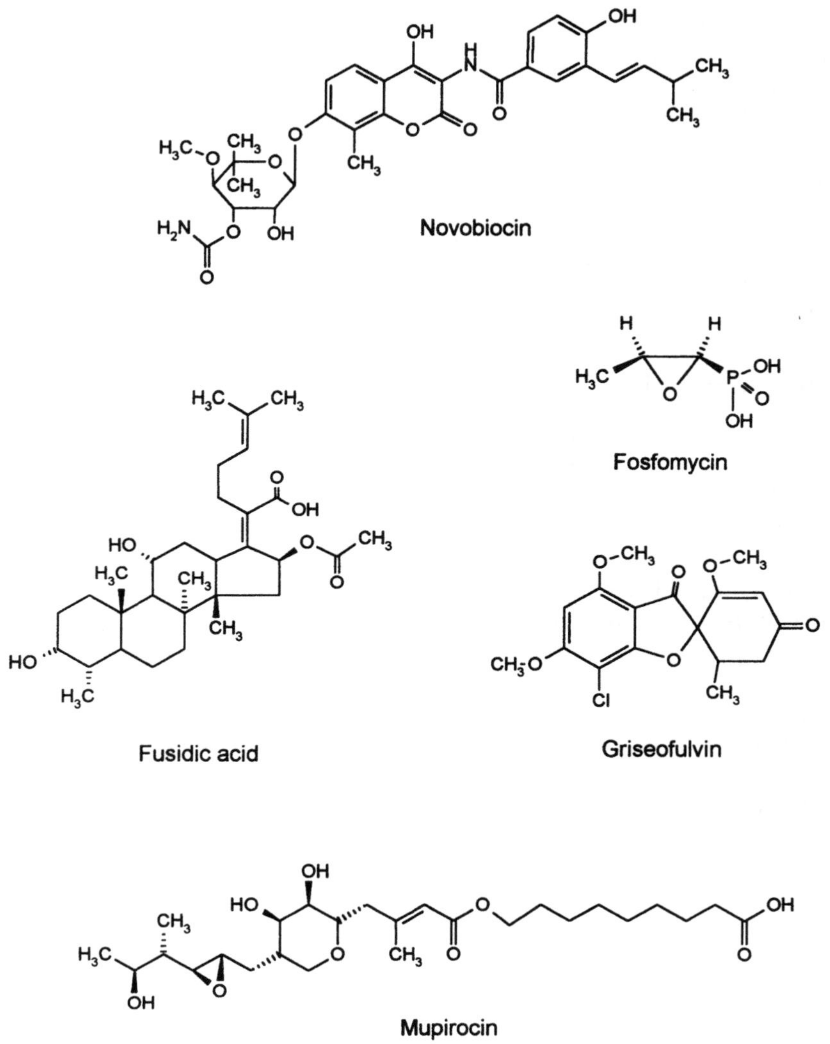

Figure 5.44. Structures of miscellaneous antibiotics.

5.10.6. Griseofulvin

Isolated and described since 1939 as a product of *Penicillium griseofulvum* metabolism, *griseofulvin* (Figure 5.44) was considered an inactive product for many years. In 1947 it was identified with the "curling factor," an antifungal activity produced by other penicillia. Only in 1958 was its clinical effectiveness in skin mycoses noted, and it was introduced into therapy.

Griseofulvin is active *in vitro* against the dermatophytes (*Microsporum, Epidermophyton, Trichophyton*), but not against other pathogenic fungi. It is effective *in vivo* in cutaneous mycoses because, when administered orally, it concentrates in the deep cutaneous layers and the keratin cells.

Numerous semisynthetic derivatives of griseofulvin have been prepared but none have activity equal to or better than that of the original product.

5.10.7. Mupirocin

The antibacterial activity of extracts of *Pseudomonas fluorescens* has been known since the beginning of the century and crude preparations have been occasionally used in therapy. The active substance, *pseudomonic acid A*, was isolated only in the 1970s, and more recently developed as a topical antibiotic with the name *mupirocin* (Figure 5.44). It is active against a broad spectrum of gram-positive bacteria, such as staphylococci resistant to several antibiotics and streptococci, but less active against enterococci. Among the gram-negative bacteria, *Haemophilus, Neisseria, Branhamella,* and *Bordetella* are susceptible. Resistant strains may be selected with low frequency from cultures *in vitro*.

Mupirocin is not active systemically as it is rapidly metabolized to an inactive product, monic acid. Topically, it is used in skin infections and for the eradication of staphylococci (normally localized in nasal fossae) in healthy carriers.

Chapter 6

Biosynthesis and Genetics of Antibiotic Production

The study of biochemical and genetic aspects of antibiotic production is of great interest because it casts light on the mechanisms of cellular differentiation. It is also important for two practical reasons: the improvement of production yields and the possibility of directing the fermentation process toward products with desired characteristics.

This chapter, like Chapter 3, has been organized using three approaches: 1) methodological, 2) systematic, and 3) exemplary.

The basic methods of study of secondary metabolism in microorganisms are first presented so that the reader can understand the actual techniques used to elucidate the biosynthetic pathways. Despite the widely different structures of antibiotics, their biosyntheses are classified into a limited number of pathways related to primary metabolism. The major pathways are described through examples of the biosynthesis of industrially important antibiotics. The most relevant findings recently obtained by molecular genetic techniques are also briefly reported.

Antibiotics are, by definition, relatively small molecules produced by microbial metabolism. Apart from their antimicrobial action, they differ from the ordinary microbial metabolites (or *primary metabolites* such as amino acids, coenzymes, and nucleotides) in that they have no apparent function in the producer organism's metabolism, and each is produced by a small number of species or strains. In this respect they belong to the larger class of compounds termed *secondary metabolites*. The expression secondary metabolite was introduced by Bu'Lock in the early 1960s (the term had been previously used by plant physiologists to indicate plant alkaloids) to indicate microbial metabolites found as the products of differentiation in restricted taxonomic groups, and not essential for cell metabolism. Typical examples are pigments, protease inhibitors, and fungal toxins. In recent times, several secondary metabolites devoid of antimicrobial activity but with other interesting biological properties have been isolated.

The study of the biosynthesis of antibiotics, or in more general terms of secondary metabolites, consists of identifying the sequence of reactions by which primary metabolites are converted into the final molecule, the mechanisms of regulation, and the governing genes' organization.

6.1. Methods of Study

The complete elucidation of a biosynthetic process ideally comprises:

1. Identification of the "building blocks," the primary metabolites from which the molecule is made
2. Isolation of intermediates of the pathway, whose structures may suggest a reasonable hypothesis as to the sequence of reactions
3. Identification of the enzymes that catalyze the single reactions
4. Identification of the governing genes and determination of their sequence and organization

It is then of great interest to identify the mechanisms and the factors that regulate antibiotic production and the way these interfere with enzyme synthesis or activity.

Although the experiments related to the different steps could ideally be performed in the above-indicated order, in practice this rarely happens. Nowadays, it is common to obtain relevant information through genetic methods before any biochemical evidence is available.

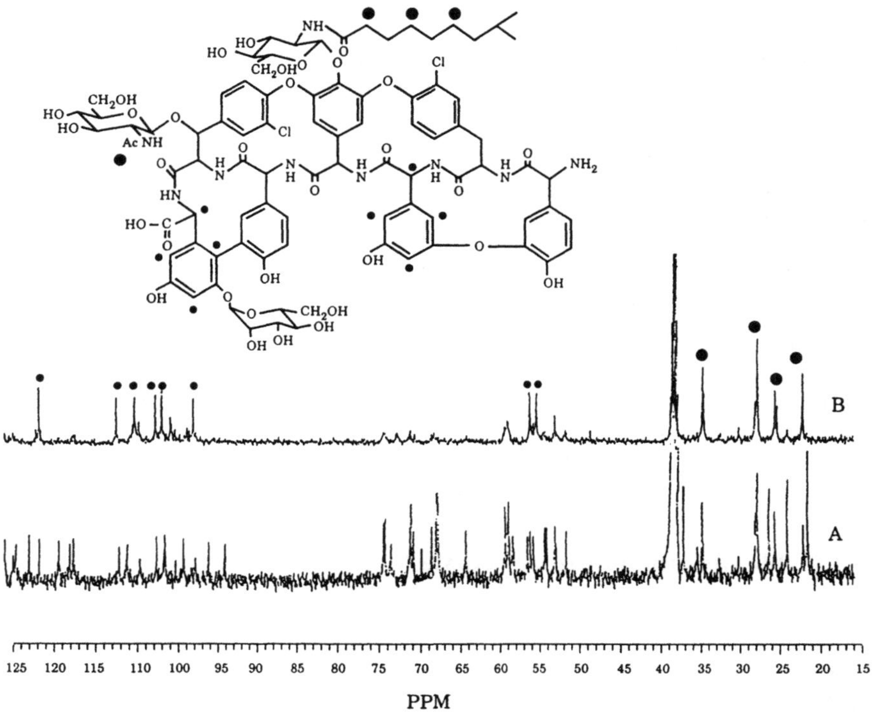

Figure 6.1. Incorporation of [2–^{13}C]acetate into teicoplanin. (A) ^{13}C-NMR spectrum (δ 20–125 ppm zone) of teicoplanin obtained by normal fermentation; (B) the same spectrum of teicoplanin obtained by fermentation in the presence of [2–^{13}C]acetate. It can be seen that acetate is the precursor of the dihydroxyphenylglycine moieties (small dots) and of the fatty acid chain (large dots).

6.1.1. Tracer Techniques

To identify the building blocks, presumptive precursors of the antibiotic, labeled with an isotope such as ^{14}C, ^{3}H, or ^{13}C, are added to cultures of the producing organism, preferably at the end of the growth phase. After an appropriate incubation period, the antibiotic is extracted, purified, and the incorporation of the isotope determined. When a radioactive label is used, it is often necessary to degrade chemically the molecule in order to establish in which atoms the radioactivity has been incorporated. When ^{13}C-labeled precursors are used, incorporation into the single atoms is revealed by the nuclear magnetic resonance (NMR) spectrum of the product. With this technique the single carbon atoms of a molecule are shown as peaks in the resonance spectrum, whose height is roughly proportional to the content of ^{13}C. The extent of label incorporation is given by the ratio of the labeled molecule's peaks to those of the natural product (^{13}C has a natural abundance of 1.1%). An example is reported in Figure 6.1.

It is sometimes useful to use precursors labeled on two adjacent atoms. A high-resolution analysis of the multiplicity of the resonance spectrum peaks and of their coupling constants can reveal whether the two atoms are linked together throughout the metabolic process or whether they are separated and incorporated through different paths.

NMR study with deuterium-labeled precursors can also give important information. Deuterium, in contrast to hydrogen, does not give a resonance signal in proton NMR. For the study of the mechanism of biosynthetic reactions, deuterium is specifically substituted for one or more hydrogen atoms at the reactive site of a biosynthetic intermediate, which is then converted into the final molecule by cultures of the producing strain. Proton NMR of the final molecule reveals whether the atom has been retained or exchanged during the enzymatic reaction.

6.1.2. Identification of Intermediate Metabolites

A common procedure suitable for identifying intermediates of the biosynthetic pathway is based on the isolation of mutants unable to perform one of the biosynthetic reactions. These particular mutants, dubbed "blocked mutants," often cause the substrate of the blocked reaction to accumulate in the medium. They can be obtained by submitting the producing strain to mutagenic treatment, and isolating clones that have lost the producing ability. These are grown two at a time in one flask (a system called co-fermentation) and the fermentation broth is assayed for antibiotic production. Strains that do not produce when they are grown singly but do produce when grown together are mutants blocked in two different steps of the biosynthetic pathway: the inability of one to synthesize an intermediate is compensated by the ability of the other to accumulate it.

The accumulated product can be isolated and identified. To verify that it is indeed an intermediate of the biosynthetic pathway, the original strain is assessed for conversion of this molecule into the final product. This can be done, for instance, by adding the intermediate to a suspension of cells of the producing strain and, after a few hours, determining the amount of antibiotic produced in comparison with a control cell suspension.

6.1.3. Identification of the Enzymes

It is sometimes possible to obtain information on a biosynthetic pathway by comparing enzymatic activities in producing and nonproducing variants of a strain. The presence of an active enzyme in a producer strain and its absence in a blocked mutant is strong suggestive evidence of that enzyme's involvement in the biosynthetic process.

When tracer techniques and isolation of intermediates have given sufficient information on the biosynthetic sequence, it is possible to study the single reactions from an enzymatic point of view. This aspect is investigated using classical biochemistry techniques: the conversion of an intermediate is studied in cell-free systems, and, when feasible, the enzyme catalyzing the reaction is purified and the reaction mechanism studied.

When a family of closely related antibiotics is produced, the substrate specificity of the biosynthetic enzymes can clarify the substrate–product relationship between two metabolites.

6.1.4. Genetics and Recombinant DNA Techniques

The understanding of the biosynthetic and regulatory mechanisms of secondary metabolite production has made great advances in the last few years as a result of techniques such as cloning and analysis of the genes involved in these processes.

Initial studies have shown that, particularly in actinomycetes, the biosynthetic genes are generally clustered on the chromosome of the producing organisms. Further studies have shown that these clusters normally include also regulatory genes and genes conferring self-resistance.

Ideally, to identify the biosynthetic genes a producer organism's genomic DNA library is constructed in a suitable vector, and the DNA fragments are transferred into a nonproducing organism. Transformed clones are then selected and examined for their producing ability. In practice, however, this simple scheme very seldom works, since the biosynthetic pathways are governed by many genes (a biosynthetic cluster may comprise over 80 kilobases) that must be transferred together and properly expressed to obtain antibiotic production. On the other hand, the amino acid sequence of biosynthetic enzymes is known for only a few antibiotics and thus, in the majority of the cases, it is not possible to use the reverse genetic technique, i.e., to synthesize, on the basis of the amino acid sequence, polynucleotide chains to be used as probes to identify the corresponding genes. Two different strategies have been adopted that often give good results:

1. Transfer of a DNA library from an antibiotic-producing organism (generally an actinomycetes) to an organism (*Streptomyces lividans*

is the species of choice) sensitive to the antibiotic, and selection of the resistant clones. From these the gene coding for self-resistance is then isolated, and used as the starting point to isolate the flanking structural genes.
2. Hybridization with DNA probes derived from genes governing similar biosynthetic pathways, taking advantage of the fact that genes coding for analogous functions in similar biosynthetic pathways normally show a high degree of homology.

When a biosynthetic gene has been identified, comparison of the nucleotide sequence with those available in data banks often reveals the nature and the function of the gene product, i.e., its enzymatic or regulatory activity. Moreover, the deduced sequence of the enzymes may help in understanding the mechanism of the biosynthetic reactions.

6.2. Biosynthetic Reactions and Pathways

The enzymatic reactions leading to the synthesis of antibiotics are not basically different from those of intermediary metabolism leading to the production of primary metabolites. In fact, it is reasonable to hypothesize that the enzymes now deputed to the synthesis of special metabolites evolved from those of general metabolism. The biosynthetic pathways by which antibiotics are formed can be classified into two large categories:

1. *Reactions modifying a primary metabolite*. These are analogous to the reactions involved in the synthesis or catabolism of primary metabolites, such as amino acids or nucleotides. The product of these reactions can be either an antibiotic or a moiety of a more complex antibiotic. In primary metabolism there are examples of molecules made by two or more units, as in the synthesis of coenzymes such as folic acid or coenzyme A.
2. *Processes of polymerization of similar metabolites*. By these processes a complex key intermediate is normally built, then further modified to give the final active molecule. We can consider four classes of antibiotics derived from polymerization processes:
    a. Antibiotics derived from polyketide synthesis, a polymerization mechanism similar to that of fatty acid synthesis
    b. Peptide antibiotics, derived from condensation of amino acids through a process similar to polyketide synthesis

c. Isoprenoid antibiotics, derived from condensation of isoprene units
d. Aminoglycoside antibiotics derived from the condensation of sugar units, as in polysaccharide synthesis

6.2.1. Relations between Primary and Secondary Metabolism

The production in large amounts of a secondary metabolite implies considerable variations in the cell's primary metabolism that must provide for the supply of starting material and energy in the form of ATP and reduced coenzymes. Therefore, it is not surprising that, when comparing producing (particularly high-producing mutants) and nonproducing strains, substantial differences are found in the level of enzymes not directly involved in antibiotic biosynthesis.

In general it should be remembered that the production of secondary metabolites is a form of differentiation, and as such implies a metabolic switch, consisting of repression of many genes and activation of others.

6.2.2. Synthesis of "Families" of Antibiotics

Often a microbial strain produces several metabolites, very similar in their chemical structure and biological activity, constituting a "family" of antibiotics. The production of families of compounds is not limited to antibiotics but is a general characteristic of secondary metabolism, and has been attributed to an apparent lack of substrate specificity of the enzymes involved. Actually an enzyme normally recognizes a specific region of the substrate molecule. Sufficiently large molecules, very similar in their structure, presenting only small differences in regions not spatially related to the site of enzymatic action, can be recognized as substrates in that reaction. The production of families of antibiotics is the result of the following series of events:

1. A key intermediate is synthesized through one of the processes mentioned above.
2. This intermediate is modified by common metabolic reactions, such as methylation, hydroxylation, etc.
3. The same metabolite can be the substrate of two or more of these reactions with formation of two or more products that can in turn be further transformed, giving rise to a *metabolic tree*.
4. Because of the apparent lack of substrate specificity, the same metabolite can be produced through two different sequences of reactions, giving rise to a *metabolic net*.

6.3. Transformation of Primary Metabolites into Antibiotics or Antibiotic Moieties

6.3.1. Biosynthetic Pathways Related to Amino Acid Metabolism

Several antibiotics, or moieties that are part of larger antibiotic molecules, originate from intermediates of amino acid synthesis or from products of amino acid catabolism. A few examples are reported here.

The biosynthesis of chloramphenicol, a product of *Streptomyces venezuelae*, (Figure 6.2), has been elucidated by isolating all the intermediates from blocked mutants of the producing strain. The carbon skeleton of the molecule derives from chorismic acid, a key intermediate in the biosynthesis of aromatic amino acids, that is converted into *p*-aminophenylpyruvic acid rather than into *p*-hydroxyphenylpyruvic acid as in the synthesis of tyrosine. The subsequent reaction is a transamination, analogous to the last step of tyrosine biosynthesis, giving rise to *p*-aminophenylalanine. Noteworthy among the other reactions of the pathway are the formation of the nitro group by oxidation of an amino group and the formation of dichloroacetic acid by chlorination of acetoacetyl-coenzyme A. An enzyme that possibly catalyzes this last reaction has been identified and found to be similar to catalases present in several microbial species.

A key intermediate in the biosynthesis of ansamycin antibiotics and of mitomycin is 3-amino-5-hydroxybenzoic acid (Figure 6.3). The unusual characteristic of this product is the presence of an amino group in *meta* position to the carboxyl group of benzoic acid (in primary metabolism only *para*- or *ortho*-aminobenzoic acids are known). The biosynthesis of this intermediate is an interesting example of how a minor modification in the biosynthetic pathway of primary metabolites can lead to unexpected molecules of secondary metabolism. In fact, 3-amino-5-hydroxybenzoic acid is synthesized through a series of reactions closely paralleling those involved in the synthesis of shikimic acid, a well-known intermediate in the biosynthetic pathway of the aromatic amino acids (Figure 6.3). The determinant difference is the conversion of a carbonyl into an amino group at the first (or possibly the second) step of the pathway.

An example of an antibiotic moiety derived from catabolism of amino acids is the alkylprolyl moiety of lincomycin. As depicted in Figure 6.4, this metabolite originates from tyrosine, through a series of reactions similar to some extent to those of melanin formation. In both cases tyrosine is first hydroxylated at the aromatic ring to give dihydroxyphenylalanine, and the alanyl side chain is then cyclized with formation of a pyrrolidine (or pyrroline) ring.

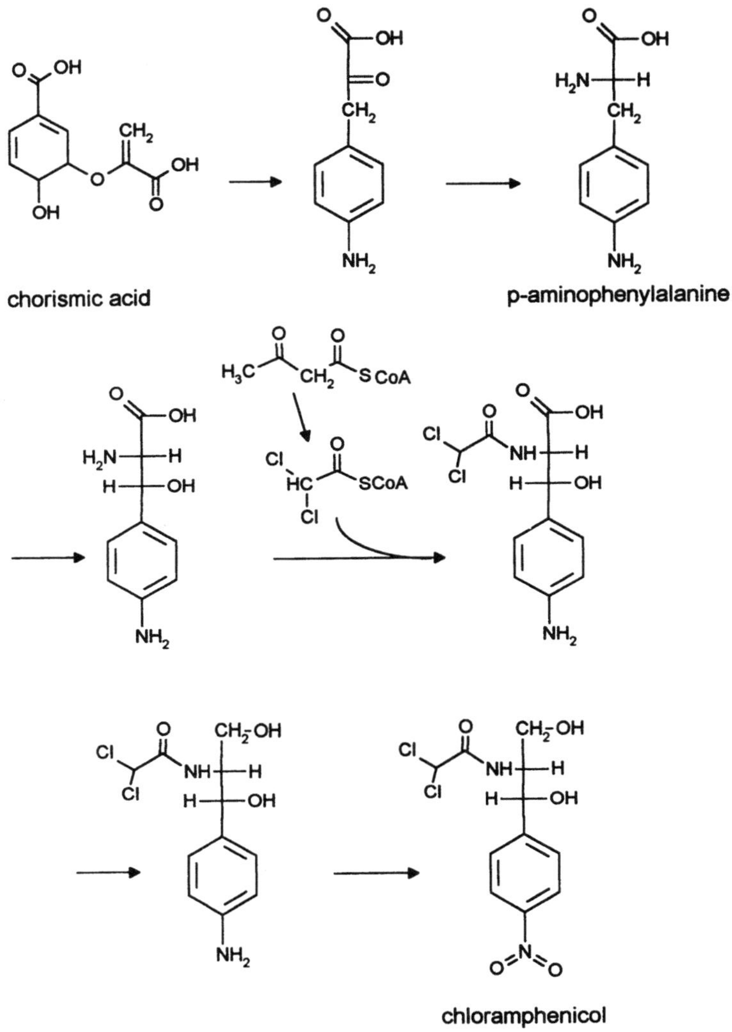

Figure 6.2. Biosynthesis of chloramphenicol.

A reaction typical of secondary metabolism is the β-hydroxylation of amino acids. We have already seen that β-hydroxylation of *p*-aminophenylalanine is one of the first reactions in chloramphenicol biosynthesis. β-Hydroxy amino acids are found as components of several antibiotic molecules, and *p*-hydroxyphenylglycine, present in glycopeptide antibiotics and nocardicins, derives from tyrosine through a series

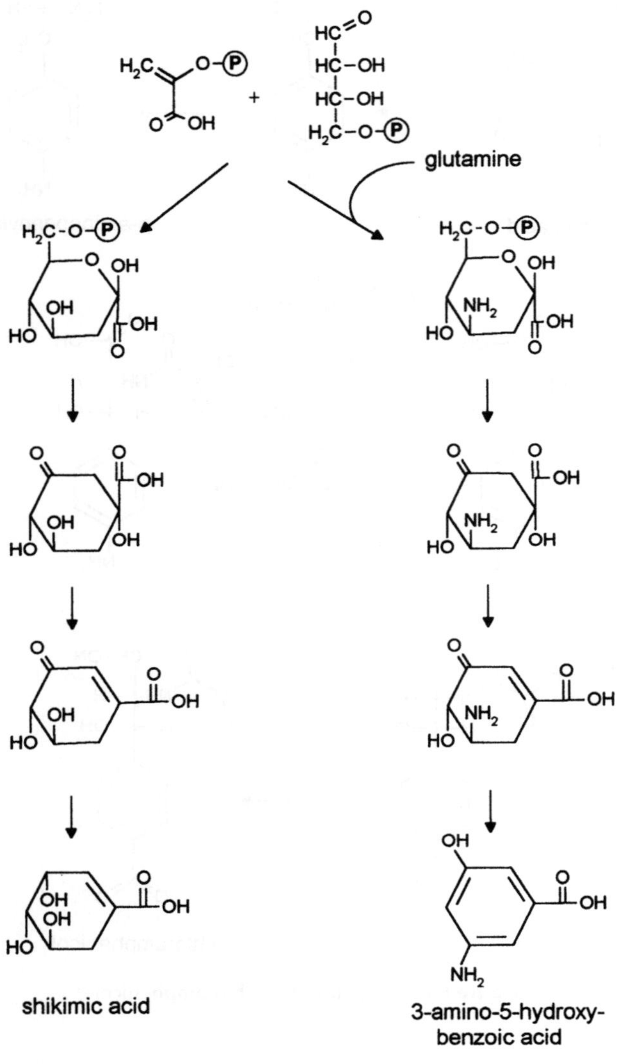

Figure 6.3. Biosynthesis of shikimic acid and comparison with the presumptive biosynthetic pathway of 3-amino-5-hydroxybenzoic acid, an intermediate in several antibiotic biosyntheses.

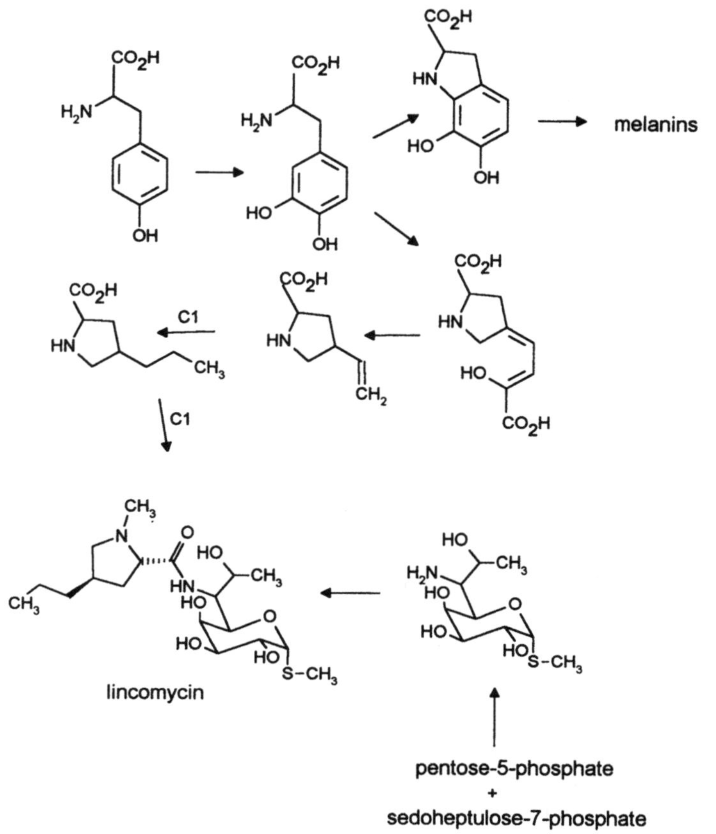

Figure 6.4. Origin of melanins and of the prolyl moiety of lincomycin from tyrosine.

of reactions initiated by β-hydroxylation of this amino acid. A further example is the recently elucidated biosynthesis of clavulanic acid (Figure 6.5) where, however, hydroxylation occurs after enclosure of the arginine α-amino group into a β-lactam ring.

6.3.2. Biosynthetic Pathways Related to Nucleotide Metabolism

Antibiotics having a nucleoside structure are synthesized either through pathways related to nucleotide synthesis or by modification of common nucleosides.

Only a few examples of the first biosynthetic pattern are known. Worth mentioning is the synthesis of antibiotic PA 399 (5, 6-dihydro-

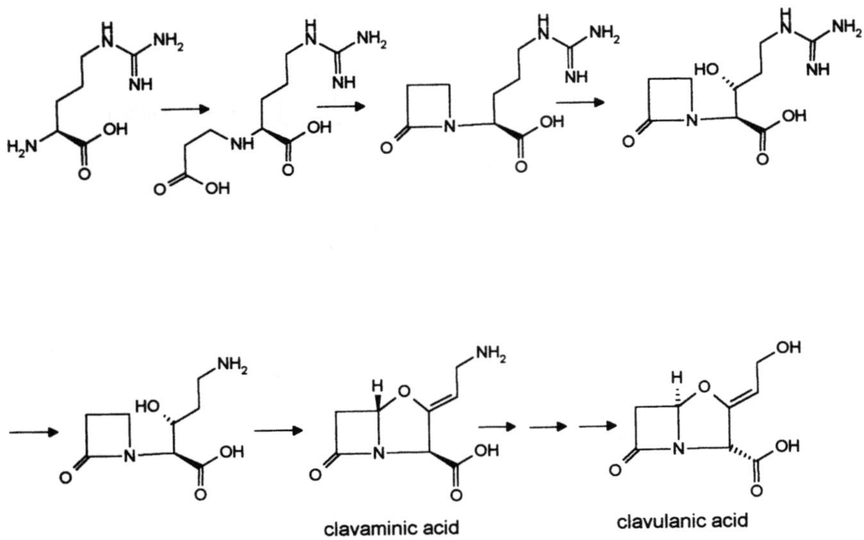

Figure 6.5. Biosynthetic pathway of clavulanic acid.

azathymidine), which closely parallels the *de novo* synthesis of pyrimidine nucleotides. Besides *de novo* synthesis, nucleotides can be formed through the so-called "salvage pathway," consisting of the ribosylation of exogenous purine or pyrimidine bases. Some antibiotics are synthesized in a similar way, such as psicofuranine, where the sugar psicose, rather than ribose, is the glycosylating agent of adenine.

Several antibiotics have been reported that derive from modifications of common nucleotides or nucleosides. An example is vidarabine (ara-A), synthesized by epimerization of the adenosine 2′-hydroxyl group. The reaction involves oxidation of the hydroxyl to a keto group, tautomerization to the enol form, and reduction of the double bond (Figure 6.6). More complex is the biosynthesis of puromycin, a product of *Streptomyces alboniger* (Figure 6.7). The first step of this pathway is the conversion of adenosine into 3′-amino-3′-deoxyadenosine, to which tyrosine is then linked by an amide bond. The sequence of the subsequent reactions has been clarified by identification of the genes involved in the biosynthesis. It should be noted that these reactions include acetylation of the tyrosine nitrogen and removal of the acetyl group as the last step in the sequence. The acetylation is believed to inactivate the otherwise active intermediates and thus protect the producing organism.

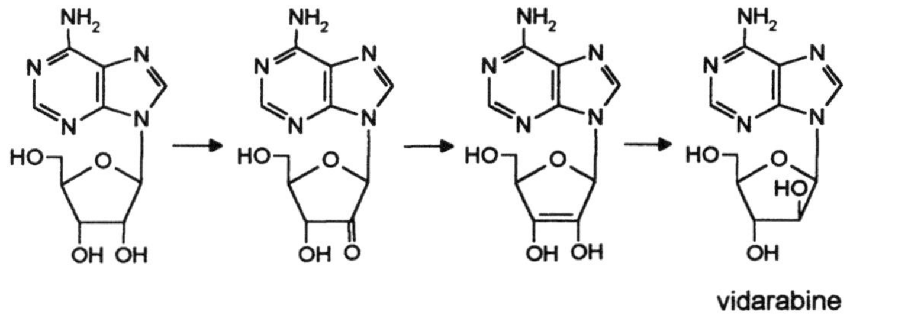

Figure 6.6. Conversion of adenosine into vidarabine.

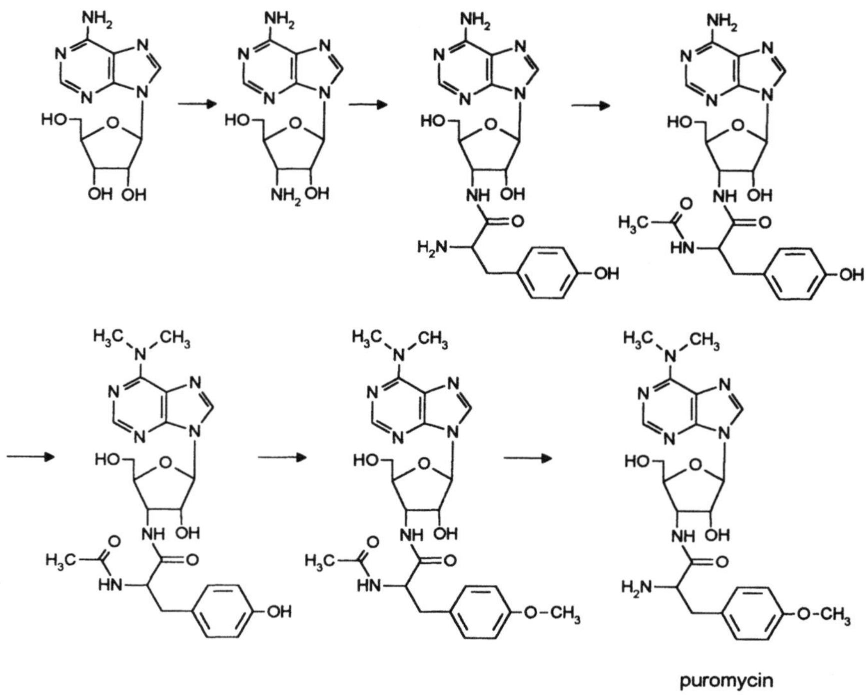

Figure 6.7. Biosynthesis of puromycin.

6.3.3. Biosynthetic Pathways Related to Coenzyme Metabolism

The pteridine moieties of folic acid and riboflavin are synthesized by a series of reactions initiated by removal, as formate, of carbon 8 of guanosine triphosphate. A family of nucleoside antibiotics, including among others tubercidin, toyocamycin, and cadeguomycin, are characterized by the presence of a pyrrolopyrimidine moiety, i.e., either a 7-deaza-adenine or a 7-deaza-guanine ring. The first reaction of the common biosynthetic pathway of these antibiotics is the removal of carbon 8 of GTP, catalyzed by GTP-formylhydrolase. The similarity of the biosynthesis of these metabolites with pteridine biosynthesis is enhanced by the observation that ribose is the donor of the three carbon atoms needed to complete both the pteridine and the antibiotic molecules (Figure 6.8).

Nicotinic acid, a biosynthetic precursor of the coenzyme nicotinamide adenine dinucleotide, is synthesized through two different pathways, depending on the organism. In mammals and *Neurospora* it derives from tryptophan, through the intermediate 3-hydroxyanthra-

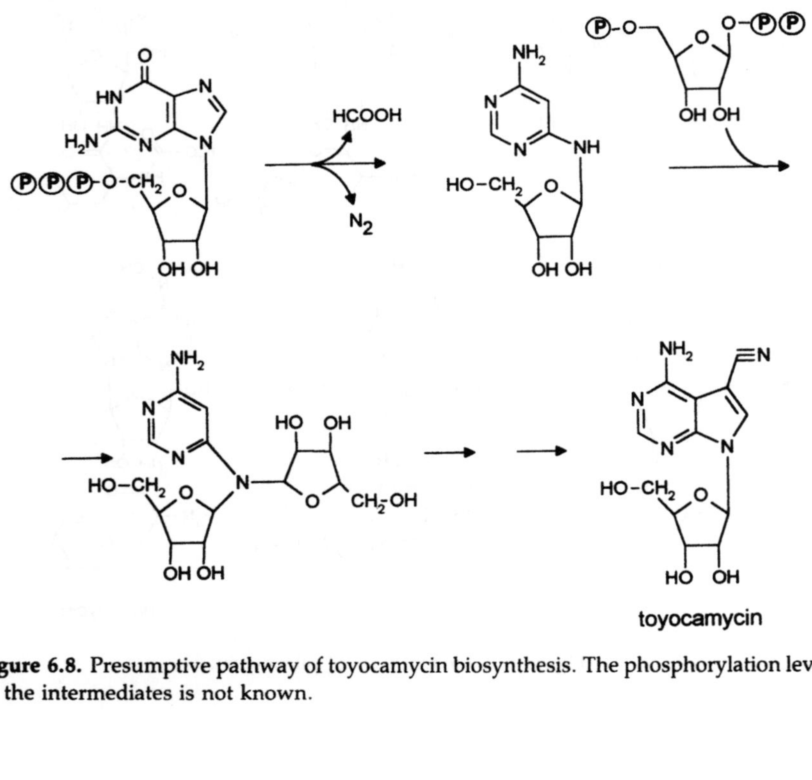

Figure 6.8. Presumptive pathway of toyocamycin biosynthesis. The phosphorylation level of the intermediates is not known.

nilic acid. In most bacteria and plants it is synthesized by condensation of aspartic acid with a three-carbon unit, such as glycerol or a related metabolite of the Krebs cycle.

The depsipeptide antibiotic pyridomycin, a product of *Streptomyces pyridomyceticus*, contains two pyrimidine moieties. These appear to be synthesized by a pathway similar to that of nicotinic acid in bacteria, as studies with radiolabeled precursors have demonstrated that both derive from condensation of aspartic acid and a three-carbon unit, probably glyceraldehyde.

A pathway related to the synthesis of nicotinic acid in mammals and *Neurospora* is involved in the biosynthesis of the aromatic core of actinomycins. This tricyclic ring is formed by condensation of two units of 3-hydroxy-4-methylanthranilic acid, which derives from tryptophan through 3-hydroxykynurenine, an intermediate of nicotinic acid biosynthesis in *Neurospora* (Figure 6.9).

6.4. Antibiotics Derived from Polyketide Synthesis

A large number of antibiotics are synthesized through a process called polyketide synthesis, which is closely related to the synthesis of fatty acids in primary metabolism.

Fatty acid biosynthesis is a polymerization process, catalyzed by fatty acid synthase (FAS). In vertebrates and yeasts FAS is a large multifunctional enzyme containing two subunits (type I FAS), whereas in bacteria and in higher plants it is a multienzyme complex, that can be dissociated into up to seven distinct enzymes (type II FAS). The polymerization process has been studied in detail, and its essential steps can be summarized as follows.

1. Acetate, the initiator molecule, and malonate, the extender unit, are linked to the synthase as thioesters: acetate to a subunit (or domain) called condensing enzyme, and malonate to the acyl carrier protein (ACP).
2. The carboxyl carbon of acetate is condensed to the methylene carbon of malonate; at the same time the free carboxyl group of malonate is eliminated as CO_2. The resulting product is acetoacetate, bound as thioester to ACP.
3. The keto group of acetoacetate is reduced to a hydroxy group. A subsequent dehydration step generates a double bond, which is then saturated by a second reduction step to give a butyrate moiety.

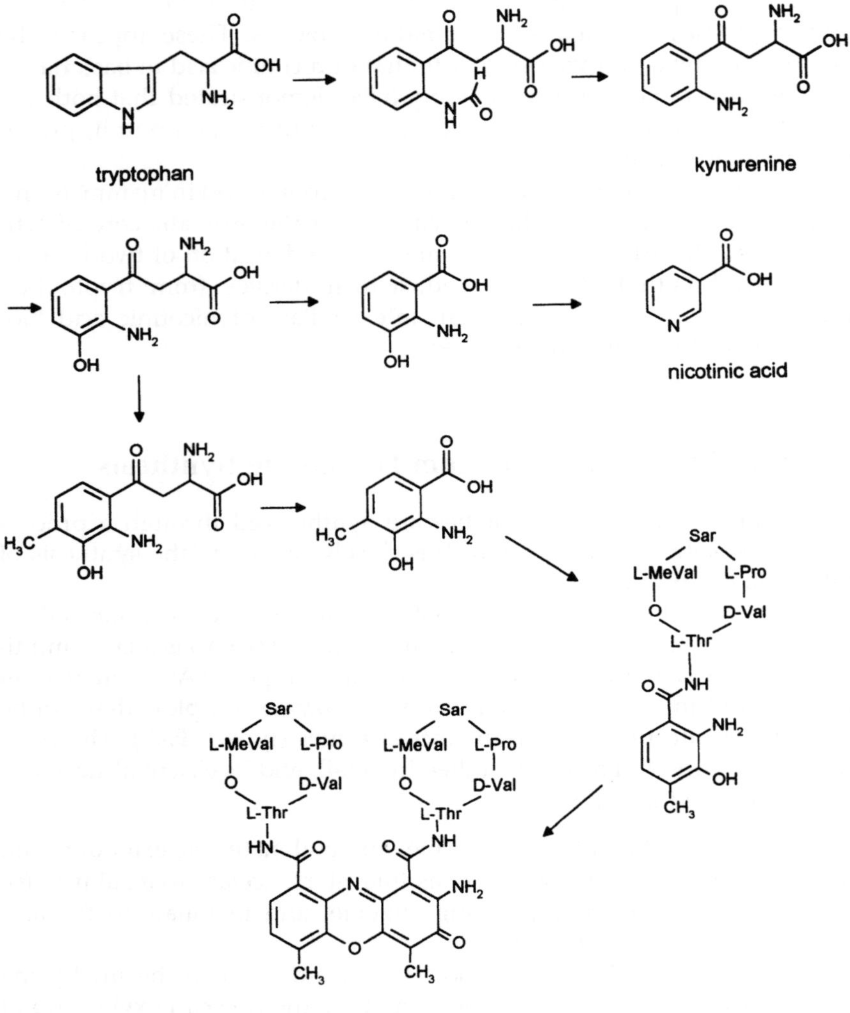

Figure 6.9. Comparison of the biosynthesis of the phenoxyazinone moiety of actinomycin D with that of nicotinic acid in *Neurospora*.

4. The growing chain is transferred to the condensing enzyme, another malonate unit is linked to the ACP, and the process is repeated until the proper length is reached. A thioesterase then releases the aliphatic acid.

In microorganisms a few variations are found in this biosynthetic pattern: in the genus *Bacillus* and in several Actinomycetales genera the initiator molecule is often a small branched aliphatic acid and branched fatty acids of 15 to 17 carbons are formed. Frequently one reduction step is omitted, and a linear unsaturated acid is produced.

The polymerization processes of polyketide antibiotic synthesis are also catalyzed by two types of synthases: type I polyketomethylene synthase (PKS I), a multifunctional enzyme, and type II polyketomethylene synthase (PKS II), a multienzyme complex. However, in contrast to the FASs either type may be found in strains belonging to the same genus.

6.4.1. Aromatic Polyketide Antibiotics

A vast number of antibiotics, produced either by actinomycetes or by filamentous fungi, have a polycyclic aromatic structure, or derive from polycyclic aromatic intermediates. Molecular genetic studies have demonstrated that, at least in *Streptomyces*, the genes governing their production encode proteins having the characteristics of type II PKS. These are similar in structure and function to type II FAS but lack the subunits deputed to the reduction of keto groups as described above at point 3 of fatty acid synthesis. The molecule resulting from the polymerization process is a polyketomethylene chain (polyketide), i.e., a linear molecule in which keto and methylene groups alternate. Such a chain is very reactive and tends to fold and form rings by aldol condensation between keto and methylene groups. According to the length of the chain and the nature of the enzymes involved, many different structures can be formed that tend to be composed of aromatic rings because of steric and energy reasons.

The best studied type II PKS is that of actinorhodin, a dimeric molecule comprising two identical tricyclic moieties, each made up of one acetate and seven malonate units (it must be pointed out that since acetate is easily converted by cells into malonate, when labeled acetate is used as a precursor the entire molecule appears derived from acetate units). Actinorhodin PKS is comprised of five polypeptides, whose amino acid sequences, as deduced from the gene sequences, correspond to those expected for the different enzymes, such as ACP, condensing enzyme, and cyclase.

The aromatic polycyclic structure is the earliest intermediate in the biosynthesis that can be isolated and identified. The biosynthetic pathways normally include a series of reactions by which it is then converted into the final molecule. As examples, the biosynthetic pathways of tetracycline and griseofulvin are reported in Figures 6.10 and 6.11, respec-

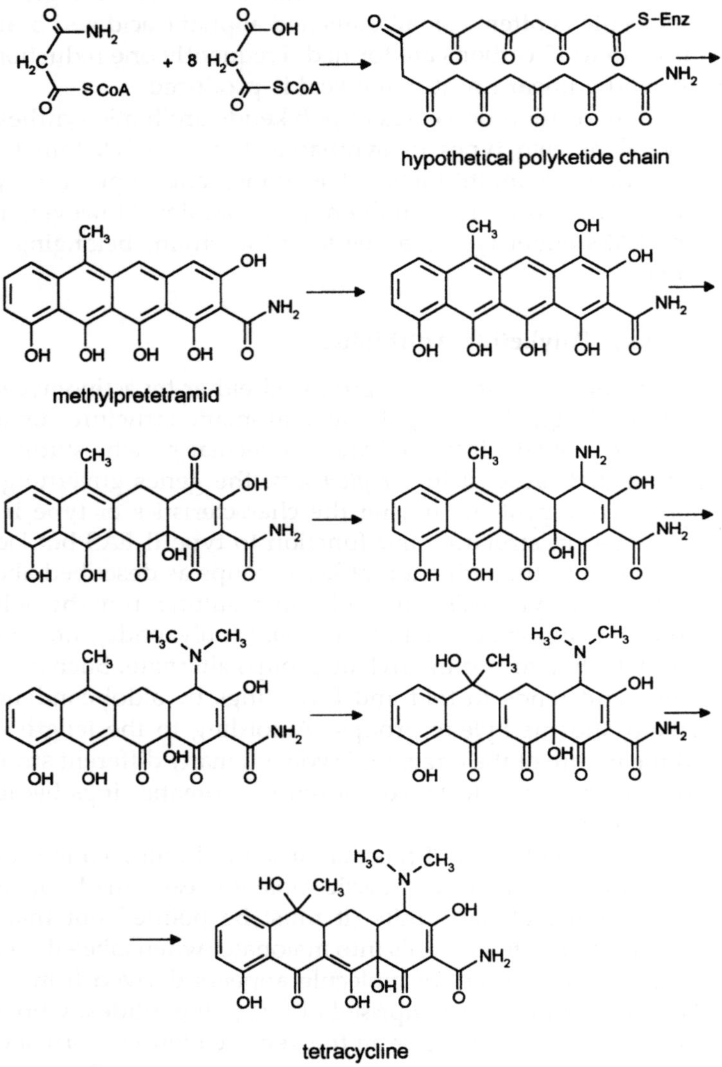

Figure 6.10. Biosynthesis of tetracycline.

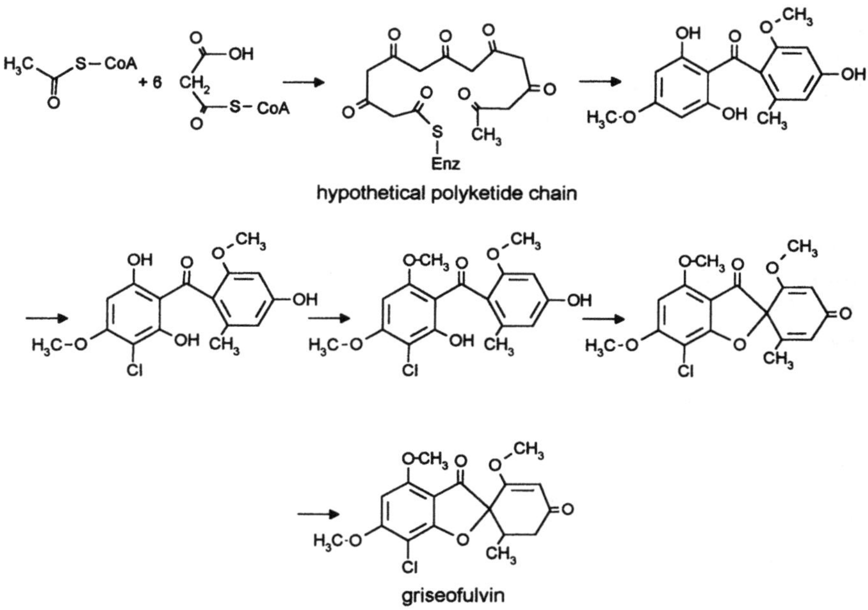

Figure 6.11. Biosynthesis of griseofulvin.

tively. It may be noted that in tetracycline biosynthesis the initiator molecule is malonamide rather than acetate.

6.4.2. Complex Polyketide Antibiotic Biosynthesis

Several classes of antibiotics are known whose biosynthesis involves the polymerization of small carboxylic acids by a mechanism related to fatty acid synthesis, but differing from the latter in many important aspects. A source of variation is the initiator molecule that frequently is propionate but sometimes consists of other aliphatic or aromatic acids. As extender molecule, methylmalonate is often found, less frequently ethylmalonate, and in rare cases propylmalonate. When methylmalonate is the extender unit the resulting chain carries methyl substituents (ethyl groups in the case of ethylmalonate) because the condensation of the growing chain always occurs at the carbon adjacent to the carboxyl group. A third important source of variation is the omission of the reduction steps, or the dehydration step, mentioned above in point 3 of fatty acid synthesis. The chain may then contain keto or hydroxy functions or double bonds. Because of this, and because of the presence of alkyl substituents, the chain does not contain a series of methylenes activated by adjacent keto groups, and therefore cannot form aromatic rings by simple dehydration. The resulting structures are then either linear molecules or large cyclic molecules closed by an amide (macrolactams), or by an ester bond (macrolactones, also called macrolides).

In the biosynthesis of fatty acids or aromatic polyketides, all elongation cycles are identical. In the synthesis of these complex polyketide antibiotics the polymerization process must be precisely programmed in order to insert the appropriate extender unit with the appropriate level of reduction at each elongation step. How this is accomplished was first elucidated by studies on organization and sequences of the erythromycin biosynthetic gene, and confirmed by similar studies on avermectin and other metabolites. A common feature among these genes is that they encode large multifunctional polypeptides, each containing one or more regions called synthase units (SU). Each synthase unit comprises domains performing the FAS-like activities necessary to complete one elongation cycle, including the selection of the extender unit, the condensation process, and, when appropriate, the total or partial reduction of the keto group. After completion of a cycle, the nascent polyketide is transferred to the adjacent synthase unit where another extender unit is added and processed, so that the order in which the SUs are aligned determines the order of the functional groups in the final molecule. The number of SUs thus determines the length of the chain. The stepwise building of the chain constituting the erythromycin backbone is illustrated in Figure 6.12.

The basic structure of the chains and their enzyme-directed folding determine to which class the final antibiotic molecule will belong. At least six important classes of antibiotics appear to be synthesized by this polymerization process. Three of these classes are characterized by a macrolactone backbone: the 14-membered and 16-membered antibacterial macrolides, the polyenes, with a ring of 26 to 38 atoms and the avermectins with a complex polycyclic structure. One class, the ansamycins, are macrolactams whose molecules comprise an aromatic ring spanned by an aliphatic chain. Two classes, the ionophoric polyethers and the elfamycins, have a linear structure.

The large number of antibiotics found in every class originate from the specificity of the other biosynthetic genes present in the different producing organisms. As examples the biosynthetic pathway of erythromycins and rifamycin B are reported in Figures 6.13 and 6.14, respectively. It should be noted that, in addition to the steps shown, erythromycin biosynthesis requires the synthesis of two sugars, mycarose and de-

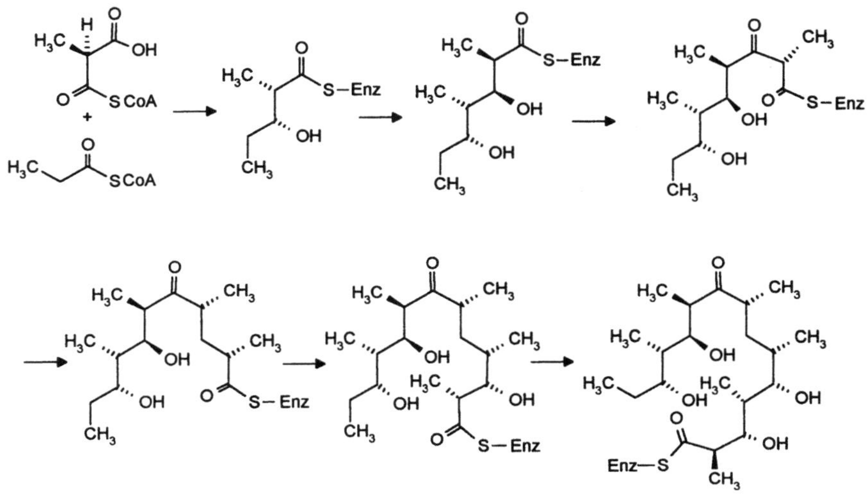

Figure 6.12. Example of a chain construction by type I polyketomethylene synthase: construction of the chain that gives rise to 6-deoxyerythronolide B in *Saccharopolyspora erythrea*.

sosamine, governed by genes that belong to the biosynthetic cluster. In the rifamycin B structure the carbon chain is interrupted by the insertion of an oxygen. This is characteristic of the rifamycin family of antibiotics, whereas in all other ansamycins the carbon chain is continuous.

6.5. Antibiotics Derived from Polymerization of Amino Acids

6.5.1. The Thiotemplate Mechanism

A large number of microorganisms, belonging to different taxonomic groups, synthesize peptide antibiotics by a process called thiotemplate mechanism. This name reflects the main characteristic of the system, by which the sequence of the amino acid residues in the antibiotic molecule is determined by the order in which the precursors are linked, as thioesters, to specific sites on one or more multifunctional enzymes that catalyze the polymerization process.

Basically the process consists of:

1. Activation of the amino acids as adenylates
2. Condensation of the amino acids to thiol groups on the enzyme
3. Stepwise polymerization

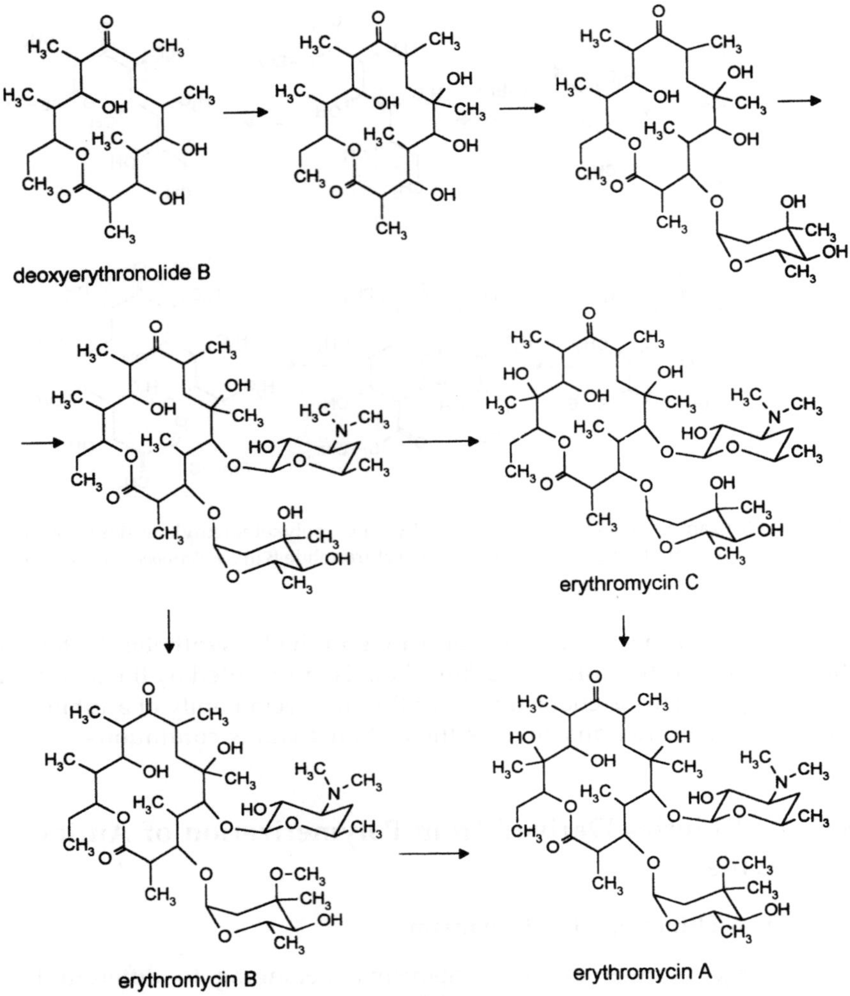

Figure 6.13. Biosynthetic reaction sequence from deoxyerythronolide B to erythromycin A.

Polymerization initiates with formation of a peptide bond between the carboxyl group of the first amino acid and the amino group of the second one. In this step and in the following ones energy is provided by breaking of the thioester bond. The carboxyl end of the dipeptide then forms a peptide bond with the amino group of the third amino acid. The

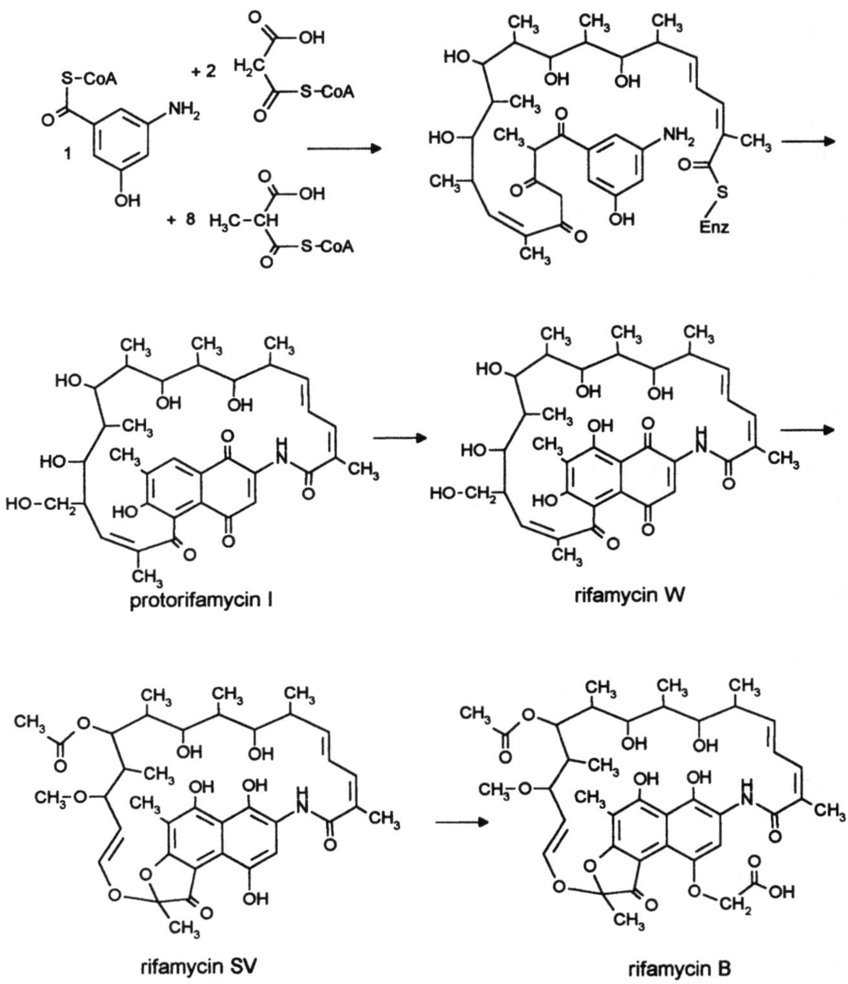

Figure 6.14. Basic steps of rifamycin B biosynthesis.

nascent chain is then condensed with the next amino acid until completion of the molecule, which is then released by a thioesterase.

There are obvious similarities between this process and the synthesis of polyketide antibiotics. Genetic studies have also revealed homologueies in the sequences of some domains of the involved genes.

The enzymatic complex that catalyzes the whole process comprises up to four multifunctional enzymes. The amino acid sequences, deduced from the sequences of the encoding genes, reveal in each enzyme the presence of modules whose number corresponds to the number of amino acids processed by that particular enzyme. Within each module, domains can be identified that are responsible for the different operations required for the completion of one elongation cycle, that is, activation of an amino acid, its esterification to a thiol group, and formation of the peptide bond. In addition, in some modules domains are present that catalyze the isomerization from L to D of the amino acid or the methylation of the amide nitrogen.

The production of peptide antibiotics is widespread in bacilli, actinomycetes, and filamentous fungi. There are, however, significant differences. Actinomycetes and fungi produce, besides peptides, depsipeptides, molecules in which amide bonds alternate with ester bonds, since in addition to amino acids, hydroxy acids are used as building blocks. In bacilli frequently the peptide chain is initiated by an acyl chain, and methylated amides are not found. As in polyketide antibiotic synthesis, the basic peptide structure is then subjected to further modifications. The best-studied biosynthetic pathways of peptide antibiotics, both from biochemical and genetic points of view, are those of gramicidin S and of β-lactam antibiotics.

The scheme of gramicidin S biosynthesis is reported in Figure 6.15. An enzyme, GS1 of relative weight 130,000, activates and isomerizes phenylalanine, the initiator molecule. A second enzyme GS2 (relative weight about 500,000) activates and carries as thioester the other four amino acids that compose the molecule. By the process described above a pentapeptide is synthesized, two molecules of which are then assembled head-to-tail to complete antibiotic formation.

The biosynthetic pathway of penicillins and cephalosporins also begins with amino acid polymerization by the thiotemplate mechanism. The responsible multienzyme, ACV synthase, has been purified from both β-lactam-producing actinomycetes and fungi. It activates α-aminoadipic acid (at its δ-carboxyl group), cysteine, and valine; binds the activated amino acids as thioester; performs the inversion of configuration of L- to D-valine; and polymerizes the amino acids into the tripeptide δ-aminoadipyl-cysteinyl-D-valine. The tripeptide is then cyclized by the enzyme isopenicillin N synthase with formation of isopenicillin N, the common intermediate in penicillin, cephalosporin, and cephamycin biosynthesis (Figure 6.16). At this point the biosynthetic pathways of penicillins and cephalosporins diverge. As shown in Figure 6.17, isopenicillin N, depending on the producer organism, is either transacy-

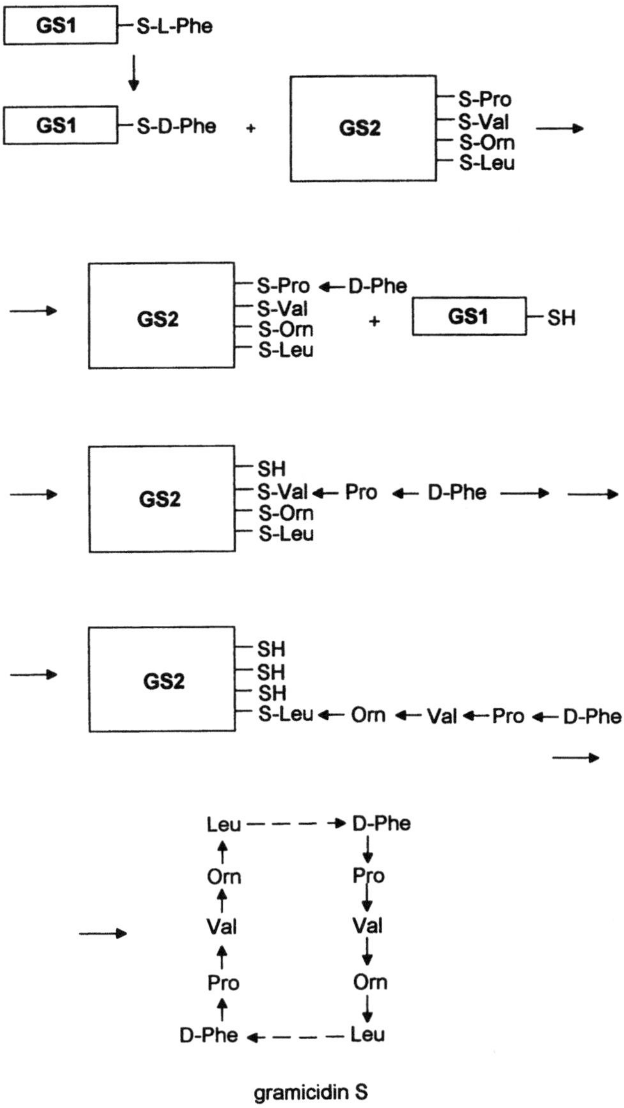

Figure 6.15. Biosynthesis of gramicidin S.

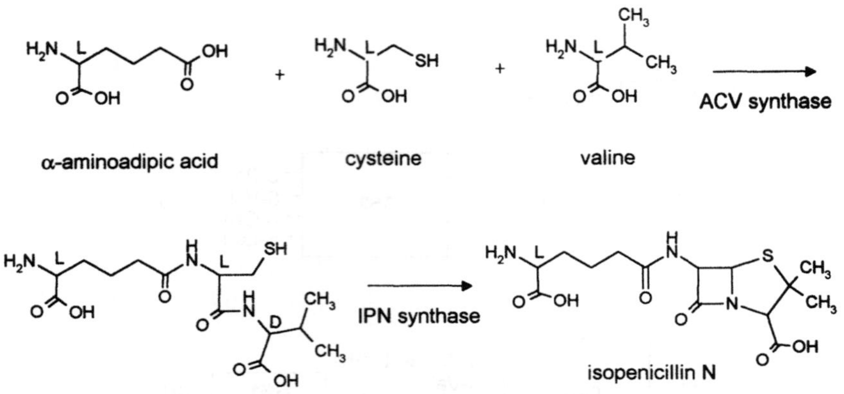

Figure 6.16. Biosynthesis of isopenicillin N, the common precursor of penicillins, cephalosporins, and cephamycins.

lated to penicillin G (or other penicillins according to the availability of acid precursors), or isomerized to penicillin N, which is subjected to ring expansion and other modifications leading to cephalosporins and cephamycins.

6.5.2. Ribosomally Synthesized Antibiotics

A small number of antibiotics are made by the general transcription and translation system of protein synthesis. These are normally larger than the other peptide antibiotics and are characterized by the presence in their molecule of lanthionine moieties, a feature from which the name lantibiotics was derived. Lantibiotics are produced by gram-positive eubacteria and actinomycetes. Those produced by eubacteria are usually larger, comprising from 20 to 34 amino acids whereas those from actinomycetes are invariably composed of 19 residues. In addition to lanthionine or methyllanthionine, other nonprotein amino acids are present in lantibiotics. However, biochemical and genetic studies have demonstrated that these derive from posttranslational modifications of ribosomally synthesized peptides, called prelantibiotics. These consist of a sequence (the prolantibiotic) that is subsequently transformed into the final antibiotic molecule and a leader peptide that is cleaved in the maturation process. The chemical modifications mainly consist of dehydration reactions; the lanthionine moieties, for instance, are formed by loss of water between cysteine and serine (or threonine) residues.

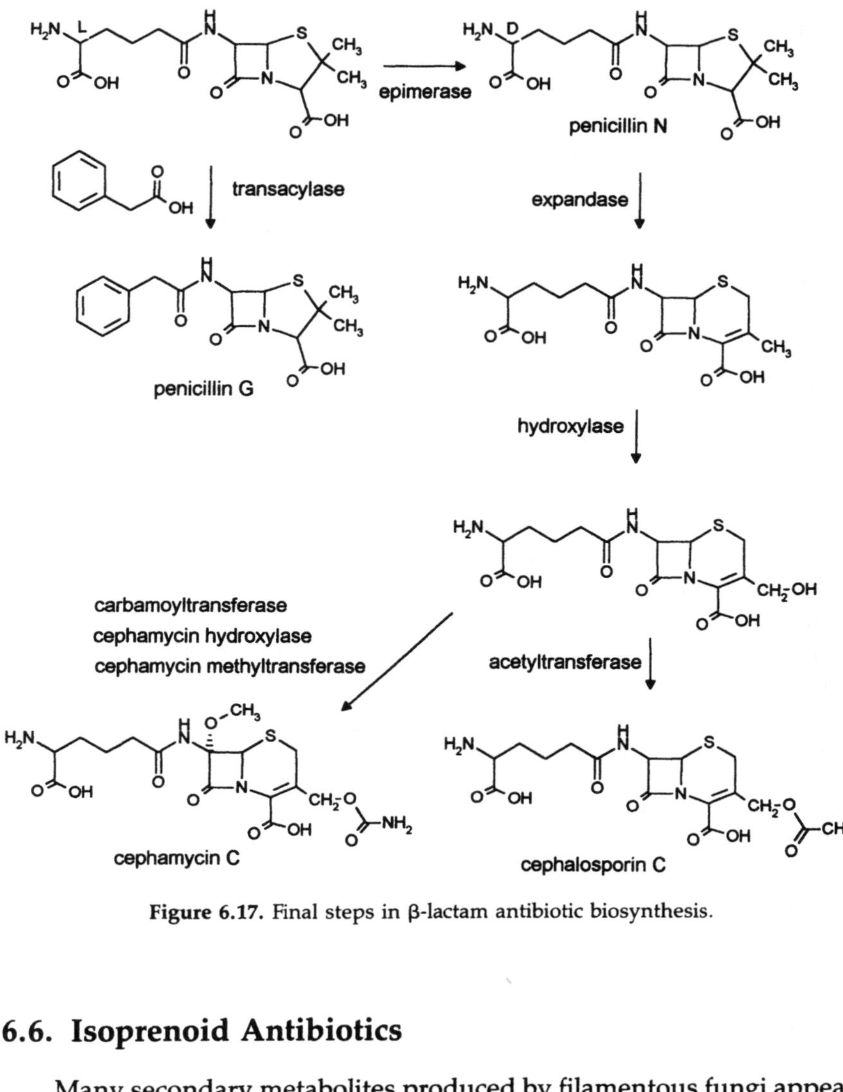

Figure 6.17. Final steps in β-lactam antibiotic biosynthesis.

6.6. Isoprenoid Antibiotics

Many secondary metabolites produced by filamentous fungi appear to be derived from the condensation of isoprene units. The biosynthetic mechanism by which these are produced is quite similar to that of the carotenoid or sterol synthesis, consisting of polymerization of isopentenyl-pyrophosphate units. Derivatives frequently found are the sesquiterpenes, made by the condensation of three units, the diterpenes, composed of four units, and the sterols. The latter are derivatives of squalene, an intermediate biosynthesized from head-to-head condensation of two molecules of farnesyl pyrophosphate.

The great variety of structures observed reflects both the different ways in which the basic chain can fold, and a number of rearrangements and other reactions that can deeply modify the original molecule. In spite of the large number of metabolites of this class that have been isolated, only one antibiotic, fusidic acid, has been found to be clinically useful. Its biosynthesis appears to be closely analogous to that of fungal membrane's sterols.

6.7. Oligosaccharide Antibiotics

Although glycosylated antibiotics are frequently found, metabolites entirely composed of sugar moieties are not very common. An important class is that of aminoglycoside antibiotics, which are actually pseudo-saccharides, being composed of sugar units and an aminocyclitol, i.e., a cyclic polyalcohol bearing amino functions. From a biosynthetic point of view aminoglycosides are derived entirely through extensive modification of common sugars, such as glucose or glucosamine. Two patterns may be distinguished with regard to these modifications: normal sugars may first be converted into biosynthetic intermediates which are then assembled, or they may first be assembled and then modified. There is no rigid distinction between the two cases; in streptomycin biosynthesis, for instance, there are both preassembly and postassembly sugar modifications.

In primary metabolism, polysaccharides are formed by stepwise condensation of sugars activated as nucleoside diphosphates at the anomeric carbon. The conversion of one sugar into another normally also occurs after a similar activation. The available evidence indicates that in aminoglycoside biosynthesis the sugar precursors are similarly converted into their nucleoside diphosphate derivatives before modification or polymerization.

6.7.1. Biosynthesis of Aminocyclitols

The majority of the therapeutically useful aminoglycoside antibiotics have 2-deoxystreptamine as their characteristic aminocyclitol. In streptomycin the aminocyclitol is streptidine. Although they both derive from glucose and show structural similarity, these two molecules are synthesized through different pathways.

D-*myo*-inositol, a common metabolite derived from glucose cycliza-

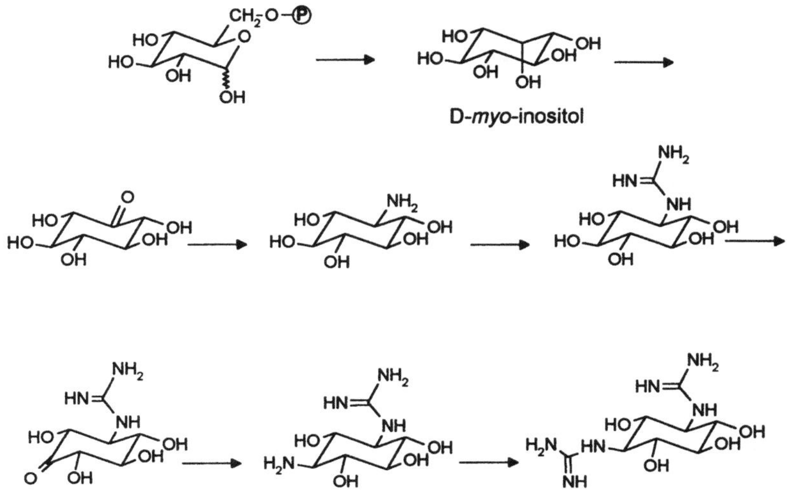

Figure 6.18. Biosynthetic pathway of streptidine. For simplicity the phosphorylation and dephosphorylation steps have been omitted.

tion, is the starting material for streptidine biosynthesis. The hydroxyl group on the carbon deriving from carbon 5 of glucose is oxidized to a keto group, with formation of *scyllo*-inosose, which is then converted to *scyllo*-inosamine by transamination. After phosphorylation of the hydroxyl γ to the amino group, the latter is converted into an amidino group by transamidination with arginine. The molecule is then dephosphorylated and an identical series of reactions is performed starting from oxidation of the hydroxyl β to the amidino group (Figure 6.18).

In 2-deoxystreptamine biosynthesis, glucose 6-phosphate is converted into 2-deoxy-*scyllo*-inosose by a complex reaction, involving several intermediate steps, closely related to the dehydroquinate synthase reaction in the aromatic amino acid biosynthetic pathway. The final molecule is then made from this intermediate by: (1) transamination of the keto group, (2) oxidation of the hydroxyl group adjacent to the methylene, and (3) transamination of the formed keto group (Figure 6.19).

6.7.2. Examples of Aminoglycoside Antibiotic Biosynthesis

Streptomycin is composed of three moieties, the aminocyclitol streptidine and the sugars streptose and *N*-methyl-L-glucosamine. The first step in the assembly of the molecule is addition of dihydro-

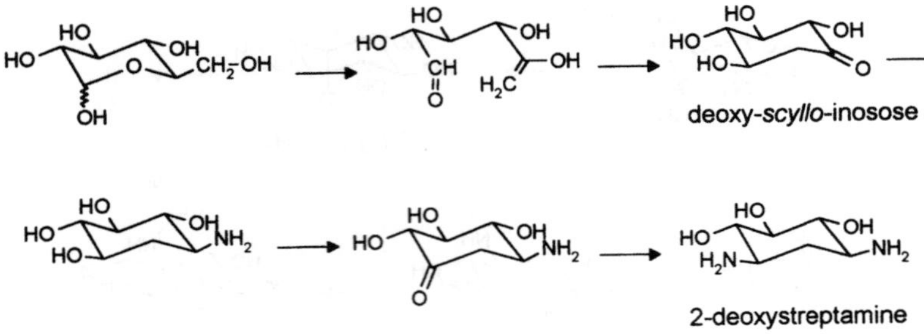

Figure 6.19. Biosynthesis of 2-deoxystreptamine.

streptose, activated as deoxythymidine diphosphate, to streptidine phosphate (Figure 6.20). dTDP-dihydrostreptose is synthesized from dTDP-glucose which is first dehydrated and then epimerized to dTDP-4-keto-l-rhamnose, a common intermediate in several antibiotic sugar syntheses. A ring-contraction reaction, catalyzed by the enzyme dTDP-dihydrostreptose synthase, produces dTDP-dihydrostreptose, in which carbon 3 of ketorhamnose becomes the exocyclic carbon.

The next assembly step is addition of *N*-methyl-L-glucosamine to the pseudo-disaccharide. The details of the biosynthesis of *N*-methyl-L-glucosamine are not known. The starting precursor is D-glucosamine, probably activated as uridine diphosphate, which is methylated and converted into the actual intermediate by a series of reactions that must involve the isomerization of four chiral centers. The final reactions are the oxidation of the exo-methoxyl group of dihydrostreptose to aldehyde and the dephosphorylation of streptidine (Figure 6.20).

The biosynthesis of the gentamicin family of antibiotics offers an example of the pattern in which most of the sugar modifications occur after the assembly of the precursors into the basic molecule structure (Figure 6.21). The biosynthetic pathway begins with the stepwise addition of D-glucosamine and D-xylose to 2-deoxystreptamine. The hydroxyl group at 3″ of the xylose moiety is converted into a methylamino group, and carbon 4″ of the same moiety is methylated (with inversion of the hydroxyl group). The resulting molecule, gentamicin X_2, is the last common intermediate of the pathway. Several different antibiotics can be derived from it, depending on the producer organism. Relevant products are sisomicin, gentamicin C_{1a}, and sagamicin, all resulting from modifications of the glucosamine moiety. *Micromonospora inyoensis* produces sisomicin from gentamicin X_2 through conversion of the 6′-hydroxyl group into an amino group and elimination, by dehydration, of the 4′-hydroxyl group. In *Micromonospora purpurea*, a further reaction reduces sisomicin to the excreted product, gentamicin C_{1a}, whereas in

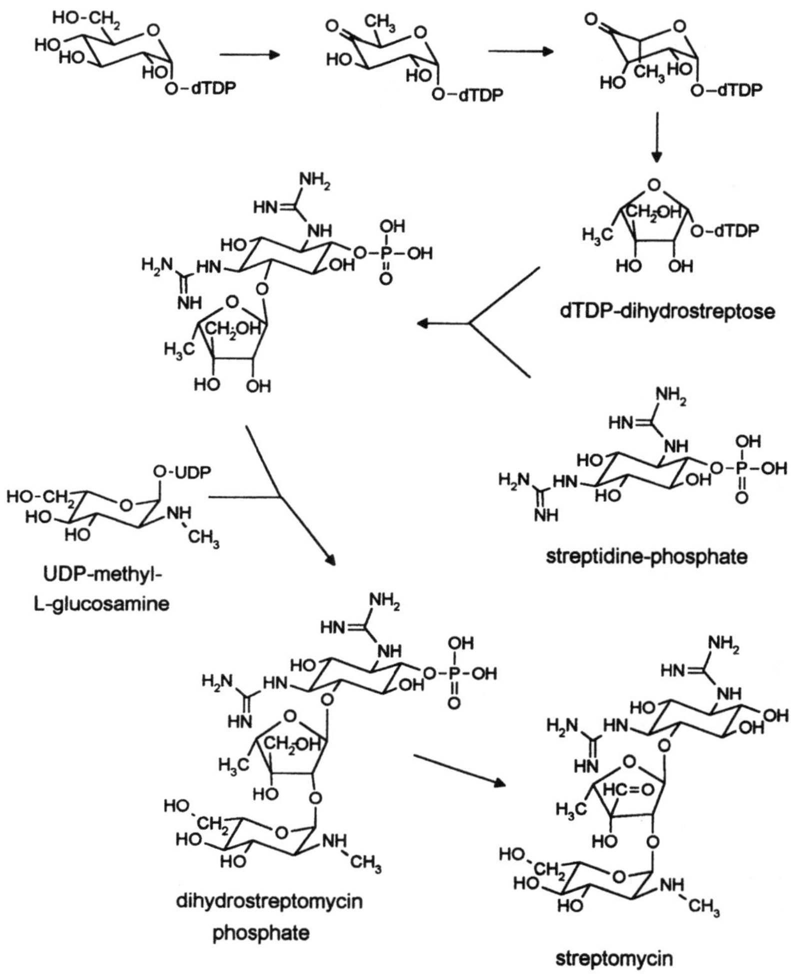

Figure 6.20. Conversion of dTDP-glucose into dTDP-dihydrostreptose, and biosynthesis of streptomycin.

Micromonospora sagamiensis the latter is methylated at the 6′-amine to give sagamicin, the final product of this strain fermentation.

6.8. Genetics of Antibiotic Production

Studies on the genetic aspects of antibiotic production are today actively pursued in several laboratories, and new results are continu-

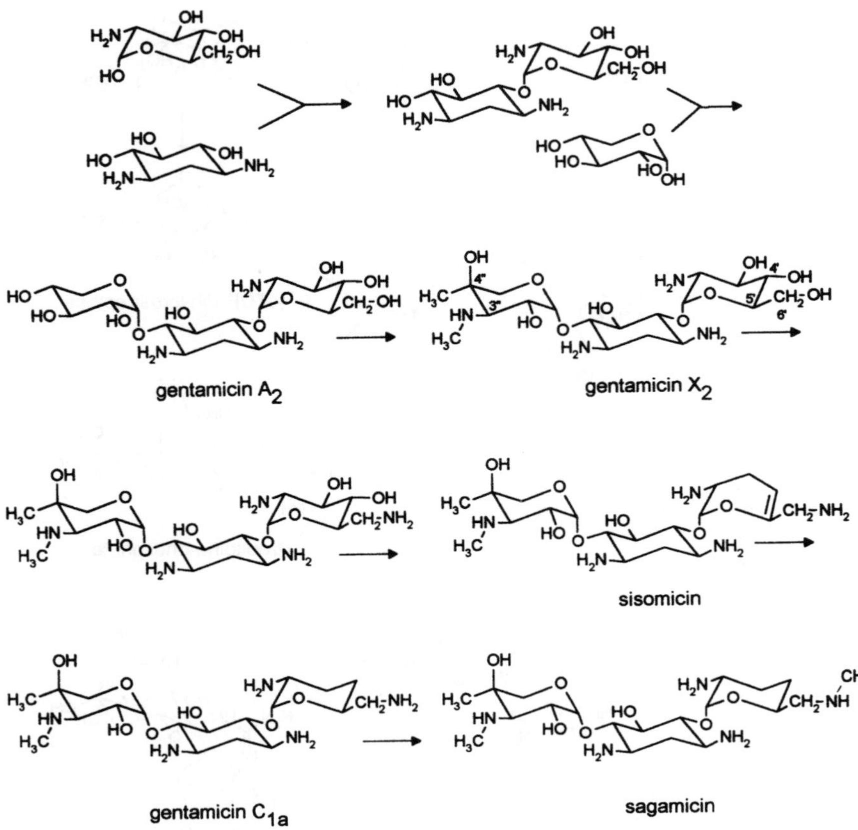

Figure 6.21. Biosynthetic pathway leading to some aminoglycoside antibiotics of the gentamicin family.

ously published that contribute to the understanding of biosynthetic pathways and their regulation. A decisive step in the advancement of these studies has been the development of a procedure for the transformation of streptomycetes (which lack a system of natural competence, i.e., are unable to take up exogenous DNA) based on the formation and regeneration of protoplasts. Although this method is not of general application, and so far only exceptionally could be extended to other actinomycetes, it has provided the means to clone and identify the genes of several important antibiotics.

An important observation emerging from the available information is that the antibiotic biosynthetic genes appear clustered, i.e., are physically linked in a segment of DNA that can span, according to the number and size of the genes, up to about 100 kilobases. It has also been observed that often regulatory genes and genes conferring self-resistance are part of the biosynthetic cluster.

6.8.1. Regulatory Genes

The biosynthesis of antibiotics is regulated by several factors, including nutrient concentration, repression or feedback inhibition by final product, induction by autoregulators, that are briefly discussed in Chapter 8. From a genetic point of view, it is important to remember that antibiotic production is a form of differentiation. Thus, we can expect that it is regulated both by genes controlling the switch from vegetative growth to morphological differentiation, and by genes specifically governing the expression of the structural biosynthetic genes. Actually, it has been observed that inactivation of the early genes involved in morphological differentiation normally also prevents antibiotic production. For instance, in *Bacillus subtilis* transcription of the sporulation genes, as well as that of the genes involved in secondary metabolism, is totally dependent on the *spoOA* gene product. Similarly, in *Streptomyces coelicolor* expression of the *bld* genes is necessary for both aerial mycelium formation and antibiotic production. These genes are not part of the antibiotic biosynthetic cluster, and their expression is possibly affected by environmental signals, such as deprivation of a specific nutrient. In contrast, regulatory genes specific for antibiotic production are expressed independently from those regulating later steps of morphological differentiation. Some genes controlling antibiotic biosynthesis have been identified in the corresponding biosynthetic clusters, e.g., *strR*, which regulates production of amidinotransferase I in streptomycin biosynthesis, and *actII*, a positive regulator of actinorhodin biosynthesis in *S. coelicolor*.

6.8.2. Self-Resistance Genes

Antibiotic-producing organisms have developed ways of avoiding inhibition by the antibiotic they produce. Some biochemical mechanisms of this self-resistance are briefly discussed in Chapter 9. As previously mentioned, at least some of the responsible genes have been found as part of the biosynthetic cluster, although occasionally additional determinants are localized in different regions of the chromosome. Following are a few examples of the first situation.

A widespread self-protection system is the temporary inactivation of the antibiotic by *N*-acetylation or *O*-phosphorylation. In the puromycin producer *Streptomyces alboniger*, for instance, the gene *pac* determines acetylation of an intermediate, and deacylation is the last step of the biosynthetic pathway. In the gene cluster governing streptomycin biosynthesis, *strA*, *strK*, *aphD*, and *aphE* determine the phosphorylation or dephosphorylation reactions of pathway intermediates and of the final product.

Two different mechanisms of self-resistance are operative in the oxytetracycline producer *Streptomyces rimosus*. Two genes, *otrA* and *otrB*, have been cloned that independently confer resistance to oxytetracycline-sensitive *Streptomyces* species. The first determines resistance by modification of ribosomes, the target of tetracycline action. The second mediates resistance through the production of a protein that enhances antibiotic efflux from the cell.

6.8.3. Structural Biosynthetic Genes

Genes coding for enzymes catalyzing biosynthetic reactions have been cloned and sequenced for an ever-increasing number of antibiotics. Here we mention the most important ones.

The *act* and *tem* gene clusters, governing actinorhodin and tetracenomycin production, respectively, have been cloned and entirely sequenced. Similar studies have been successfully performed on the genes responsible for other aromatic polyketide antibiotics, such as oxytetracycline, and daunorubicin. The biosynthetic genes in the larger clusters for complex polyketide antibiotics have been cloned and completely sequenced in the case of erythromycin, and substantial information is known for tylosin and avermectin.

Among the aminoglycoside antibiotics, much information is available on the streptomycin biosynthetic gene cluster.

Detailed information has been published on the genes governing β-lactam production, which have been cloned from fungi such as *Penicillium*, *Aspergillus*, and *Cephalosporium*, and from the actinomycetes *Nocardia lactamdurans* and *Streptomyces clavuligerus*. Comparisons of the gene sequences and organization in these different organisms have provided insight on the evolution of these genes, and on their possible transfer from prokaryotic to eukaryotic organisms.

The amino acid sequences deduced from the biosynthetic genes are largely contributing to our understanding of the individual biosynthetic steps and the mechanistic details of several pathways. This information also provides the basis for the production of modified antibiotics through genetic manipulation. An example is the organization of the genes coding for polyketide synthase type I discussed in Section 6.4.2. The identification of modules, within the synthase units, that determine single reduction steps has allowed the production of modified erythromycins by gene inactivation. Similar studies are under way on other antibiotics in several academic and industrial laboratories.

Chapter 7

The Search for and Development of New Antibiotics

The search for *new antibiotics,* fifty years after the introduction of penicillin into clinical use, continues to be very active. In fact, there are still microorganisms against which the available drugs are not effective, and the spread of resistant strains has made some antibiotics that once were very effective practically useless. On the other hand, new antibiotics are still being found despite many decades of intensive search.

In this chapter both traditional and innovative methods of performing *screening* (i.e., the set of operations by which potential producer microorganisms are isolated and the substances they produce are identified) will be described.

Identification and evaluation of the biological properties of a new antibiotic represent only part of the research. A careful examination of its *pharmacological* and *toxicological* properties is an essential step before experimenting with it on man. Extensive *clinical experimentation* is necessary before an antibiotic can be proposed for therapy. These aspects are examined briefly in the section "Development of an Antibiotic from the Laboratory to the Clinic." In the following section the process of *productivity improvement* is discussed, a necessary step in transforming a laboratory substance into an industrial product.

7.1. The Search for New Antibiotics

After the discovery and evaluation of penicillin fifty years ago, systematic programs aimed at the discovery of new antibiotics were initiated all over the world. The results of these screening activities are the well-known antibiotics now available. Methods of research have not changed substantially during these decades: They involve isolating microbial strains from soil or other natural sources, culturing the strains, and identifying the active substances they produce.

7.1.1. Selection of Potential Antibiotic-Producing Strains

Microbial strains, putative producers of antibiotics, are selected and isolated as pure colonies. Experience has shown that by far the largest number of antibiotic-producing microorganisms belong to soil-inhabiting species. For this reason, a large number of diverse soil samples are obtained and subjected to procedures for isolating microbial strains. The traditional scheme is as follows: 2–4 g of soil is well dispersed in water and allowed to settle. Portions of the supernatant are placed, after suitable dilution, on a nutrient agar medium and incubated until colonies appear. These are then isolated and cultivated as pure cultures, in flasks of complex nutrient liquid medium. The flasks are incubated for 2 or 3 days after growth has ceased, thus allowing antibiotic synthesis and secretion. The possible presence of an antibiotic in a culture broth is determined by testing the antibiotic activity with one of the methods described in Chapter 2.

7.1.2. Extraction and Purification of the Antibiotic

Once an antibiotic-producing culture has been detected, the substance responsible for the activity is extracted and purified. For this, currently available techniques for chemical extraction of natural substances are applied. It is worthwhile mentioning that these techniques have noticeably improved in the last few decades, and their description is beyond the scope of this book. Obviously, the procedures will vary for different antibiotics and will depend on the physicochemical properties of the particular substance, especially on the solubility characteristics in various solvents, on the acid–base ionization, and on the affinity for chromatographic absorbents.

7.1.3. Determination of the Novelty of the Antibiotic

It is increasingly more common during a research program to isolate new antibiotics to find products that are already known. Obviously, it is very important to establish as soon as possible whether the antibiotic is new or already known. The problem is often very difficult, for two main reasons: (1) the freshly isolated producing microorganisms synthesize very small amounts of antibiotic that must be separated and purified from large amounts of material (residues of nutrients and various metabolites) present in the culture broth and (2) the scientific literature contains descriptions of about 10,000 different antibiotics.

An antibiotic is identified by systematic comparison of its properties with those of known antibiotics. This process is called *dereplication*. The principal properties of an antibiotic useful for its identification can be classified as follows and as summarized in Table 7.1. Some of these characteristics, such as the ratio of activity toward different strains, the activity against strains resistant to other antibiotics, the chromatographic mobility, and the visible–ultraviolet absorption spectrum, can be determined also on samples containing a small amount of antibiotic together with a large amount of inert material. Therefore, these properties are especially suitable for novelty assessment of an antibiotic sub-

Table 7.1. Parameters for Identification of an Antibiotic

Microbiological parameters	
	Antibacterial spectrum
	Co-resistance spectrum
	Special effects spectra (pH, serum, ions, inoculum)
	Frequency of resistant mutants
Biological parameters	
	ED_{50}, different routes of administration, various pathogens
	LD_{50}, different routes of administration, various animals
Solubility and extractability	
	Different solvents, pH effects
	Resins (anionic, cationic, weak, strong)
Chromatographic behavior	
	Paper, thin layer, with different solvent mixtures
	Electrophoresis
	Countercurrent distribution
	Liquid chromatography at high and low pressures
Physicochemical properties	
	Spectroscopies: UV-VIS, IR, NMR, mass
	Functional analysis
	Rotation of polarized light
	Isoelectric point (electrofocusing)
Mechanism of action	
	Inhibition of synthesis of macromolecules in growing bacteria
	Inhibition of enzymatic reactions in cell-free systems
Stability	
	Light, pH, specific enzymes

stance present in crude or semipurified extracts, as they do not require extensive purification.

7.1.3.1. Microbiological Properties

These include:

1. The antimicrobial spectrum
2. The spectrum of cross-resistance
3. The spectrum of special effects, such as the effects of different test conditions (e.g., pH, presence of serum, concentration of specific ions, size of the inoculum) on the antibiotic activity
4. The frequency of emergence of resistant mutants

7.1.3.2. Biological Properties

The principal ones and those easiest to measure are the ED_{50} (see Section 7.2.2) and the LD_{50} (see Section 7.2.4) in the mouse, determined after various routes of administration. More important than the absolute values is the ratio of ED_{50} to LD_{50}, which is independent of the degree of purity of the antibiotic sample being tested as long as this contains only one biologically active substance. This is because the contaminants are normally not toxic, as experience demonstrates.

7.1.3.3. Phase Distribution Properties

These include *solubility characteristics* in different solvents (including the effect of pH on solubility), the absorption onto different resins (anionic, cationic), which gives an indication of the molecular charge, and the *chromatographic characteristics* on different supports and in different solvent mixtures.

With classical paper or thin-layer chromatography the antibiotic is ordinarily visualized by UV light and revealed by microbiological assay. This latter technique, of historical importance but still used, consists of overlaying a plate or paper strip chromatogram with an agar medium containing the test microorganism. After a suitable incubation time, the agar becomes turbid as a result of bacterial growth except in the zones over the chromatogram in which the antibiotic is present and diffuses into the agar. Tetrazolium blue or other classical indicators may be used to enhance the contrast between the growth and no-growth zones. In addition to chromatography, electrophoretic separation, which is usually faster, can also be used.

Most used today are HPLC techniques by which the substances present in the sample are separately visualized as peaks of UV light absorption. The active substance is identified by collecting various fractions and determining their antimicrobial activity.

It should be mentioned that both classical and HPLC chromatographies can be used not only for qualitative and quantitative analyses but also for preparative purposes, i.e., for the isolation of small quantities of pure and homogeneous products (this technique is very important when dealing with antibiotic complexes).

7.1.3.4. Physicochemical Properties

The most useful of these are the absorption spectra in the ultraviolet, visible, and infrared regions, the nuclear magnetic resonance spectra for hydrogen and carbon, and the mass spectra using the various ionization techniques. Among the latter, the so-called soft ionization techniques (Field Desorption and Fast Atom Bombardment) allow the determination of the exact molecular mass and exact elemental composition, the latter by the use of a high-resolution spectrometer. Another important diagnostic element is the stability of the antibiotic activity toward various factors, such as light, pH, and enzymes.

7.1.3.5. Mechanism of Action

The determination, even if approximate, of the mechanism of action has essentially two functions: an *identification* function, which is of great help to the chemist, as it enables the program of novelty determination to be focused on the comparison with known antibiotics with the same mechanism of action, and a *prediction* function, as it gives some clues about the selectivity of action and thus the potential intrinsic toxicity of the test product.

7.1.4. Originality and Probability of Success

The screening procedures that have been described do not differ substantially, with the exception of some technical innovations, from those employed in the past. Nowadays, it has become ever more difficult to discover, develop, and introduce new antibiotics into medical practice for two major reasons. First, a new antibiotic can be marketed only if it has properties superior to those of already existing products. Second, the criteria for admission of new products established by the regulatory authorities of different countries have become ever more stringent, excluding from the market many new products that would have been accepted in previous years. As a consequence, the probability of a successful outcome from a screening program becomes smaller and smaller. There is thus a continuous effort to devise technical or conceptual innovations that might be introduced into a screening program to increase the probability of success. The proposals advanced by the various experts deal with three categories of factors:

1. Quantitative factors
2. Qualitative factors
3. Organizational factors

7.1.4.1. Quantitative Factors

The influence of these factors is straightforward. All other factors being equal, the higher the number of fermentation broths examined and of tests to which they are submitted, the greater is the probability of finding a useful substance. The quantitative dimensions of a program are limited by obvious cost considerations. However, the number of strains isolated and tested in a given time can be considerably increased by suitable application of automated techniques and microtesting procedures, assisted by computers (see Section 2.4).

7.1.4.2. Qualitative Factors

These can be subdivided into three classes:

1. Screening new producer organisms. The widely held opinion is that the probability of discovering a new antibiotic is very small if we continue to isolate and test the same microorganisms that have been screened over the last 40 years. This opinion is based on the frustrating, unfortunately frequent, experience of rediscovering an already known antibiotic. At least 15 million organisms have been isolated, almost 10 thousand different antibiotics have been described and for the most part completely characterized, and about 200–300 new ones are described each year. However, many experts are convinced that not all types of antibiotics that exist in nature have been discovered, and that the chance of finding new useful chemical structures would be greater if we would examine unusual, rare microorganisms not studied in the past. This opinion is based on practical experience and on the idea that the production of microbial secondary metabolites is a form of differentiation characteristic of the producing species.

Two problems arise from this proposition, one theoretical and one practical. The theoretical one may be stated in the following terms: "how far apart on the taxonomic scale, or perhaps, more exactly, on the evolutionary scale, must two groups of microorganisms be in order that their secondary metabolites will be sufficiently different?" It is already known that secondary metabolites, including those with antibiotic activity, are quite different in the bacilli, actinomycetes, and lower fungi (the three large groups of antibiotic producers), and that all three groups have been subjected to extensive screening. Therefore, it is necessary either to examine different classes of organisms or to select, within these three groups, subgroups (genera) that have not been studied in detail. But the unusual microorganisms are exactly those that have not been extensively examined because they are difficult to isolate and/or grow.

In the last few years, several new techniques and selection programs have been described that are suitable for isolating from soil uncommon microorganisms such as myxobacteria or rare genera of actinomycetes.

2. Varying the culture. In every screening program, the first step is isolation of microorganisms from soil followed by growth, usually in liquid medium, of the colonies to be tested. Let us assume that the colonies isolated belong to a rare genus that presumably differs genetically from the common ones. It is known that secondary metabolites are dispensable substances that may not necessarily be produced by a microorganism even when it has the genetic information to do so. Furthermore, many secondary metabolites are produced during the *idiophase* of the culture, i.e., after growth completion. In many cases their synthesis is controlled by mechanisms such as final product repression by glucose catabolites, nitrogen compounds, or phosphate. Therefore, it is not sufficient to have available microorganisms with a new genome but it is essential to grow them under conditions that permit maximal expression of the genetic information.

   In general, these conditions are not those that support rapid growth, and the culture medium must be chosen accordingly. In some laboratories, the isolated strains are cultured in different media. In all cases, the incubation should last some time after growth has ceased, and in the idiophase the levels of carbon and nitrogen sources and the phosphate concentration should not be too high.

3. Use of new and original test procedures. The essential aspect for the success of a screening program is considered to be the use of

tests enabling the selection of the desired type of activity directly on fermentation broths. This allows the following work of dereplication, isolation, and evaluation of the active products to be concentrated on a much lower number of strains, with increased efficiency of the whole system.

These selective tests can be conceptually divided into three groups:

a. Tests aimed at the novelty of the product. They include the utilization in the primary screening of microorganisms resistant to most common classes of antibiotics or microorganisms against which only few antibiotics are active. Alternatively, one can take advantage of a differential activity: for example, an antibiotic selectively active against staphylococci and not against streptococci or vice versa has a good probability of being new. Another example may be when a certain activity is evident only under particular conditions of the medium, such as pH or temperature.
b. Tests revealing congeners of known products. A screening program may have as its objective the isolation of compounds similar to a known and important antibiotic. For example, tests have been envisaged for the identification of β-lactams, based on strains hypersensitive to these antibiotics. Other bases for the tests are the reversion of the activity in the presence of β-lactamases or the ability to induce the expression of these enzymes.
c. Tests based on mechanism of action. It was shown in Chapter 3 that inhibitors of some metabolic functions have a greater probability of being selective toward microorganisms and thus nontoxic. A number of tests have been studied that are able to indicate, directly in the fermentation broth, the mechanism of action of the antibiotic activity. For example, cell wall inhibitors will be active against a normal bacterial strain but not against its "L-form," i.e., against a mutant that lacks the wall itself. This fact can be exploited to set up a test system that reveals strains producing antibiotics acting on the cell wall.

Reversion tests can also be utilized. For instance, to isolate inhibitors of a given enzyme one may determine the activity of the fermentation broth against a suitable microorganism and then repeat the determination after adding a large quantity of the enzyme itself to the test system. The disappearance or a marked decrease of activity indicates the formation of a complex between the active substance and the enzyme and suggests that the enzyme is the target of the antibiotic's action.

Alternatively, cell-free tests for inhibitors of enzymatic activity can be easily devised as can tests demonstrating interference with the formation of complexes between hormones and their biological receptors.

Analogous concepts can be applied to the search for specific inhibitors of different organisms, such as viruses, also by means of biochemical or enzymatic tests. Apparently, there are no limitations to the creativity of researchers in this field.

7.1.4.3. Organization Factors

These determine how a research program should be established. A successful antibiotic must have the following minimal characteristics: it must be new, it must cure infections in animals, and it must not have serious toxic effect at the doses used for treatment.

Although the program described above should increase the probability of finding a new active product, it obviously cannot guarantee that every activity found will be new. Nor will it assure in any way that products active *in vitro* as inhibitors of bacterial growth will also be effective in curing experimental infections or that they will be nontoxic. Thus, these three fundamental characteristics must be demonstrated directly for any presumptive new product.

Some organization alternatives can be visualized by asking other questions. Should one first determine the effectiveness and the nontoxicity of the product, followed by determination of its novelty, or should this be done in the reverse order? In addition, since the properties that characterize and distinguish an antibiotic from all the others are numerous and must be obtained through different tests, which ones should be studied first and which later? These questions serve to illustrate that a primary screening program must be comprised of a group of subprograms using different techniques, aimed at obtaining information about one or more of the three essential properties of the product (novelty, effectiveness, nontoxicity) with the lowest cost and in the shortest time.

It is important to find the correct balance between the different operations that constitute a screening program. Different laboratories may have different attitudes, according to their objectives and experience, with respect to the proportion of work and time that should be devoted to each set of operations. The following examples are worthy of discussion.

The first dilemma is whether to screen many strains grown using one culture condition, or whether to screen a lower number of strains grown under several culture conditions. There is no universally accepted answer to this question, as demonstrated, for instance, by the fact that in some laboratories a single growth medium is used whereas in others each strain is grown in four different fermentation media. A consideration that may be used as a rule of thumb is that it may be convenient to use one condition when the strains screened belong to genera abundant in soil samples and easy to isolate. In contrast, when strains difficult to isolate or rare strains are examined it may be convenient to grow them in a variety of media.

A similar problem is related to the number of cultures examined and the number of assays in which each culture is tested. It is naturally desirable to test the cultures in as many assays as possible, but this may result in a reduction of the number of strains screened. A consideration that has to be made, in evaluating the correct balance between these two sets of operations, is that the probability of finding an active culture may be quite different for the different assays. For instance, an assay aimed at detecting inhibitors of protein synthesis, or of bacterial cell wall synthesis, will reveal, because of the large number of enzymes involved, active substances at a higher frequency than an assay for inhibitors of a single enzyme. As a consequence, one should screen a higher number of cultures when a very specific assay is employed; however, specific assays are often more laborious and time-consuming than the general ones. An often used compromise is to consider two sets of assays in a two-stage process: simple primary assays, such as those detecting an antimicrobial activity, on which all the cultures are tested, and secondary more complex ones (such as biochemical assays of enzymatic activity) on which only the cultures positive in the primary screening are examined. This may reduce by severalfold the number of complex tests to perform.

The next complex decision, which must take into consideration several factors, is the amount and type of work that should be done when an extract of an active culture has been obtained. Often an evaluation of the biological activity can be performed on a crude sample. When the results are not particularly promising it has to be decided whether to continue with the purification and related chemistry, which is normally rather laborious, or to discard the product. Possibly the outcome will depend on the objectives of the laboratory. In an industrial laboratory, where only products promising for practical application should be pursued, the work could be discontinued, whereas in an academic institution the chemical characterization of a novel natural product could be considered of interest.

7.2. Development of an Antibiotic from the Laboratory to the Clinic

Let us imagine that a screening program whose prime focus is the discovery of new antibiotics, as described in the preceding section, has produced one or more substances with interesting antibacterial activity and that we want to develop it (them) into products to be used in medical practice. This brings about a complex program, here briefly described, that evaluates in detail the pharmacological properties of the molecule.

7.2.1. Microbiological Evaluation

In addition to determining the spectrum of action of the new antibiotic against the standard laboratory strains, it is very important to assess its activity against the pathogens currently identified in clinical practice. A large number of pathogenic strains isolated from patients must be collected and the MIC_{90}s of the product being examined must be determined on these.

Other important aspects are:

1. Bactericidal activity, i.e., the MBCs on different species and the killing curves at various concentrations
2. Frequency of mutation to resistance in different species, nature of the resistance (one-step versus multistep), and possibly the biochemical mechanism by which it acts
3. Mechanism of action, which may give information on possible cross-resistance

7.2.2. Experimental Infections

One cannot go directly into clinical trials with new products having only *in vitro* information. Ethically, only products that have a high probability of being useful and whose tolerability has been ascertained in different animal species should be tested in man.

Therefore, possible candidates must be selected from among the available compounds by determining their effectiveness in curing animals infected with pathogenic bacteria, i.e., the antibiotic must cure *experimental infections*. As is always the case in new drug research, the experimental model never correlates exactly with the real clinical disease. Therefore, the information obtained from experimental infections does not have absolute predictive value, but it does increase the chance of making the correct decision, especially when effectiveness is correlated to toxicity data in the same animal species.

There are no fixed procedures suggested by health authorities to demonstrate the *in vivo* efficacy of a new antimicrobial agent. The indications by the World Health Organization simply request that experimental infection trials should allow an educated guess of the possible effects on man. Although the experimental protocols ordinarily adopted may vary according to the product and to its therapeutic activity, they are usually classified as follows.

7.2.2.1. Basic Screening Models

The most commonly used model for the evaluation of antimicrobial agents is a systemic infection in the mouse, normally a septicemia. This model is quite suitable for routine evaluation, as it consists of a single infection and a simple treatment regimen, can be performed in a short time, and presents a good reproducibility of the infection effects and of the therapy results. Other models possessing similar characteristics are meningitis and pneumonia in the mouse. The infection is commonly induced by an inoculum large enough to be lethal within a few days. The animals are treated according to a standard regimen that begins immediately or shortly after the infection time, using different antibiotic doses.

The ability of the different doses to promote survival of the infected animals is determined and expressed as the *effective dose*, ED_{50}, i.e., the *dose* (expressed in milligrams of antibiotic per kilogram of body weight of the animal) *effective in curing 50% of the infected and treated animals*. The comparison between the results obtained by different routes of administration indicates the relative efficacy of the oral or parenteral administration. The parallel treatment of noninfected animals serves as indication of the toxicity of the product being examined.

The predictivity of this model has a number of intrinsic limitations that are related to the discrepancies between the experimentally simulated illness and the true one occurring in man. These are:

1. The development of the infection is fulminating.
2. Because of the quick infection progression and the early administration of the antibiotic, the results represent prophylactic capability rather than true therapeutic efficacy.
3. The models are quite dependent on the dimension of the inoculum, a fact that may give rise to false positives or negatives.

4. The pharmacokinetic differences may be increased by the effect of a single or few doses and may influence the results markedly, thus preventing an accurate comparison among different products.

Despite these limitations, the screening models mentioned give an acceptable indication of the potential of a new antibiotic, including efficacy, route of administration, and toxicity. In practice, the early decision whether or not to develop a new antimicrobial agent is often taken on the basis of these results.

7.2.2.2. Monoparametric Models

These are experimental models by which a single indicator of antibiotic efficacy is measured, instead of the total therapeutic effect. Typical examples are the determination of the number of bacteria present after the treatment of an infection localized in a single organ, such as the heart in the case of experimental endocarditis, or the kidneys in the case of pyelonephritis. Another type of experimental infection is the so-called *topical infection*, which includes infections of skin wounds produced surgically or by burning and infections of the cornea of the rabbit's eye.

The antibiotic can be given parenterally or applied topically to the infection site. Evaluation of the effectiveness will be different for different types of infection. But the criterion for cure may be based on the differences in rates of healing of the wound in treated and untreated infected animals, or on the bacterial count after the treatment.

These models are quite similar to the screening ones described above, but from an experimental point of view they appear more complex and they are thus utilized in a subsequent stage of the product evaluation. In particular, they are useful for the evaluation of the effectiveness of products against chronic infections and, in general, they are fairly representative of the real illnesses, as infection has a slow development and the treatment scheme consists of many days of administration.

7.2.2.3. Discriminating Models

These are technically complex models that are designed to reproduce with good accuracy the beginning and the development of a given infection in man. Unlike the previous ones, in these models different efficacy parameters are measured so that one can obtain an indication whether the antibiotic is adequate for the treatment of a specific infection. They can also be used for testing the validity of new therapeutic strategies such as a new combination of antibiotics or different dosage plans. Considering their scope and their technical complexity, these models are utilized only in advanced stages of antibiotic development.

7.2.3. *In Vivo* Activity versus *in Vitro* Activity

As already mentioned, there are now about 10,000 antibiotics described, but only a small percentage of these are effective in curing experimental infections. Some of the most important reasons for *in vivo* inactivity of substances active *in vitro* are listed in Table 7.2. Among them, the two most common ones are: lack of selectivity, so that curative doses are also toxic, and pharmacokinetic properties of the substance, which may be metabolized or excreted too rapidly.

7.2.4. Toxicity

For an antibiotic to be introduced into medical practice, it must be nontoxic to the patient. This lack of toxicity is relative, since all products are toxic if given in sufficiently high doses. It is important that a reasonable margin of safety exist between the effective doses and those at which toxic signs appear. For obvious reasons, the potential toxicity of a drug in man must be extrapolated from experiments done in animals, which poses the problem of specific toxic effects in some animal species, briefly discussed later.

Table 7.2. Reasons Why Substances Active *in Vitro* May Be Inactive *in Vivo*

Microbiological
Metabolic state of bacteria (dormant state, anaerobic conditions, etc.)
Presence of nonpathogenic microbes that inactivate the antibiotic
Pharmacokinetic
The antibiotic is poorly absorbed or rapidly excreted
The antibiotic is rapidly metabolized to inactive products
The antibiotic cannot reach the bacteria located in abscesses, fibrin deposits, intracellularly, in bones, etc.
Biochemical
The biochemical environment of the infection site may antagonize the antibiotic effect (pH, antagonism by antimetabolites)
Strong binding by serum proteins can sequester the antibiotic
Toxicity
Effective doses cannot be given because they are toxic to the host

Table 7.3. Values of LD_{50} and ED_{50} for Some Antibiotics Given by Different Routes to the Mouse[a]

Antibiotic	MIC (mg/mL)	LD_{50} (mg/kg)			ED_{50} (mg/kg)	
		os	ip	iv	os	sc
Chloramphenicol	4	2460	1320	100–200	116	26
Penicillin G	0.05	5000	3490	2340	1.7	0.1
Erythromycin	0.1	2927	660	426	48	5.8
Streptomycin	10	5000	1400	120	—	1.2
Lincomycin	0.5	—	1000	214	40	30
Tetracycline	0.5	3000	200–300	160	12	6
Rifampin	0.005	770	340	585	0.12	0.11
Ampicillin	0.1	5000	3400	—	3	11

[a]The experimental infection was *Staphylococcus aureus* septicemia.

7.2.4.1. Acute Toxicity

Some indication of the toxicity of a drug is obtained by determining the so-called *acute toxicity*, which is usually expressed as the LD_{50}, the *dose* (expressed in milligrams of product per kilogram of body weight, mg/kg) *that is lethal for 50% of the animals treated*. The LD_{50} varies with the route of administration (oral, subcutaneous, intramuscular, intraperitoneal, intravenous). Table 7.3 shows the acute LD_{50} values in the mouse for some antibiotics. Today, to reduce the number of animals subjected to experiments, the LD_{50} is seldom determined. Survival of animals after a single large dose is considered as an adequate parameter of acute toxicity assessment.

7.2.4.2. Subacute and Chronic Toxicity

A more in-depth evaluation is made by experiments of subacute and chronic toxicities, which involve daily administration of different doses of the product for 1–3 or 6–24 months, respectively. These experiments are carried out in at least two animal species, of which one must be a nonrodent mammal. The aim of this type of study is to determine (1) the *maximal safe dose*, i.e., that resulting in no toxic effects, and (2) which *organs* or *functions* are most damaged by toxic doses of the product.

Furthermore, the less obvious toxic manifestations may be discovered by observing carefully during the entire treatment the animals' behavior, the consumption of food and water, body growth, and the appearance of macroscopically visible pathological changes. At specific intervals, blood chemistry determinations and various organ function tests are performed. At the end of the treatment, the animals are sacrificed and the individual organs examined both macroscopically and histologically.

It is normally possible to use data obtained in one species of animal to predict the toxicity of the product in another species, particularly in man, as there are considerable similarities between species with regard to the basic biochemical processes of cellular metabolism. There are certainly no substantial differences between different animal species in protein synthesis, DNA replication, glycolysis, oxidative phosphorylation, etc. However, some cases have been found in which the toxicity of a product is quite different in different species. With rare exceptions (precisely because it is impossible to foresee these differences, the initial administration of a drug to man is done very cautiously), the causes of these variations in toxicity, which can usually be seen more clearly during chronic treatment, lie in the different abilities of the various species to absorb, excrete, and metabolize the exogenous substances. These abilities have evolved differently in different mammalian species, usually in relation to the different dietary requirements.

Therefore, it is not surprising that an exogenous substance such as an antibiotic is absorbed or excreted at different rates in different species, and that in some species it may accumulate to a point at which it has damaging effects, especially when the antibiotic is administered repeatedly in large doses. For this reason, toxicological studies must be accompanied by comparative pharmacokinetic studies.

In the extrapolation of toxicity data from laboratory animals to man, one must remember that there are differences in the normal bacteria flora in different species. A classic example of this is the high toxicity of repeated doses of penicillin in guinea pigs, although penicillin is one of the least toxic antibiotics in other animals and in man. Penicillin kills the gram-positive flora in the guinea pig intestine, which is then colonized by gram-negative bacteria, resulting frequently in a fatal bacteremia. The intestinal flora of man is quite different from that of the guinea pig, and if the toxicity of penicillin had been tested only in the guinea pig, this product would probably have been discarded!

7.2.4.3. Teratogenesis and Genotoxicity

To ascertain that the product has no untoward effect on embryo development (teratogenesis), different doses of the product are administered to female rabbits and rats during the entire pregnancy period, and the litters are carefully examined for possible malformations.

Studies on possible genetic effects are performed both *in vitro* and *in vivo*. Mutagenicity can be assessed *in vitro* by the so-called Ames test, which consists of determining the rate of mutation induced by the product in microorganisms, and *in vivo* by several tests, such as the micronucleus test.

A more complete assessment of the possible effects of the product on fertility and reproduction is obtained by administering the drug at the time of mating to both males and females and examining the progeny to the second generation.

7.2.5. Pharmacokinetics

Pharmacokinetics is the study of the absorption, distribution, and excretion of a drug in a living animal. It also includes the study of the metabolism of the drug, i.e., the chemical transformations it undergoes in the body, usually through enzymatic reactions.

Pharmacokinetics is extremely important for determining the possible therapeutic applications of an antibiotic for two main reasons:

1. To be effective, an antibiotic must reach the infection site in concentrations greater than the MIC and remain there for a sufficiently long time. Pharmacokinetic studies tell us the concentration in the various tissues and organs as a function of time, thus permitting the prediction of therapeutic effectiveness, and give initial indications about doses and treatment intervals.
2. Comparing pharmacokinetics in different animals and in man helps to determine the animal species in which toxicity studies are more predictive of effects in man.

We must emphasize that from this point of view antibiotics are not a homogeneous group. For every antibiotic one should determine which of the species most commonly used shows pharmacokinetic and metabolic properties most similar to those of man.

The study of the pharmacokinetics of antibiotics is easier than for other drugs as their antimicrobial activity can be used as an analytical tool to determine their concentration in biological fluids and tissues. More recently, the advancement in instrumental analytical techniques has created the possibility of determining the antibiotic concentration in biological fluids and tissues on the basis of their physicochemical properties (see Section 2.3.1). However, a complete pharmacokinetic and metabolic study must include experiments with radioactively labeled antibiotics. These are usually obtained by fermentation when an available strain is a good producer, and the biosynthetic pathway or the main precursors of the antibiotic are known. After administration of the labeled compound, the amount of radioactivity present in blood and in various tissues is determined at different times, giving an exhaustive picture of the distribution of the drug and of its metabolites.

7.2.6. Clinical Studies

The aim of clinical trials is to establish the therapeutic usefulness of an antibiotic, the dose and the optimal regimen at which it should be given, and the adverse reactions it may cause.

Clinical pharmacology studies begin with the administration of single doses to healthy volunteers to ascertain that the product is well tolerated and to determine the blood levels attained at different doses. These data are essential for determining the dose to use in the initial therapeutic trials. We must keep in mind that the ED_{50}, the curative dose determined in experimentally infected animals, is not a good indication of the dose that will be effective in man, given the differences between a fulminating experimental septicemia and the slower course of normal clinical infections as well as the different pharmacokinetics. Therefore, the initial dose and schedule for therapeutic trials are usually chosen in order to obtain reasonably constant blood levels greater than the MIC during the entire treatment, especially when the antibiotic is bacteriostatic. For the bactericidal ones it is, probably, more important to obtain high peaks rather than constant levels in the blood.

Because of the numerous objective symptoms of infections in comparison with other illnesses (decrease in fever, disappearance of bacteria, etc.) the initial trials can be carried out under so-called *open* conditions, in which the physician can judge the results obtained with the product and compare them with those seen with other antibiotics. Later, the validity of the antibiotic must be determined exactly in *controlled* conditions, in which the effectiveness is compared with that of another antibiotic already in clinical use. When blind trials are performed, the products are given under a code so that there will be no unconscious bias in evaluation. For ethical reasons, in controlled trials of antibiotics one does not include the third group of comparison, that of patients treated with *placebo* (by definition a substance innocuous and ineffective).

Another aim of clinical trials is to evaluate tolerability and side effects. For these purposes, the controlled experiments yield more reliable data, because the adverse reactions are often evaluated from subjective criteria and may be related to causes other than treatment, i.e., to the infectious disease itself or to a concomitant disease.

Table 7.4 outlines the schedule for the different types of experiments

Table 7.4. Simplified Scheme of Antibiotic Development

Development of producing strain	Discovery and development of the antibiotic		Time	
Wild-type strain	Microorganisms isolated from soil		Year zero	Discovery
	Fermentation			
	Activity assay			
Studies on culture media and conditions	Extraction of the active product			
	Evaluation of: Novelty, efficacy in experimental infection, acute toxicity, absence of mutagenic properties, local tolerability			
	Patent		Year 2	
Mutagenesis and selection of high-producing strains	Subacute toxicity on two animal species			Development
	Clinical tolerability trials on healthy volunteers	Pharmacokinetics and metabolism	Phase 1	
	Efficacy trials on humans (search of effective dose)	Chronic toxicity on two species	Phase 2 Year 5	
Pilot plant fermentation studies	Clinical experimentation on humans (at least 1000 cases)		Phase 3	
	Efficacy	Side effects monitoring		
Industrial production	Application for registration		Year 10	

and the time normally needed from the initial discovery to the introduction of the antibiotic into medical practice. The time and costs of this development tend to increase continuously.

7.3. Development of an Antibiotic from the Laboratory to the Manufacturing Process

Antibiotic-producing strains originally isolated from soil usually synthesize a few tens of milligrams of product per liter of culture. To carry out all the studies necessary for evaluating the compound, several kilograms of product are needed. Moreover, if the product is introduced successfully into clinical practice, tons must be produced every year.

The cost of the product would be prohibitive were the productivity limited to that of the original strain. Obviously, one of the first steps in the development of an antibiotic is to find means to increase biosynthetic productivity. This objective may be approached in two different ways: improvement of strain productivity and improvement of fermentation conditions.

7.3.1. Improvement of the Producing Strain

This expression (the expression "development of the producing strain" is also used) refers to a set of operations that are applied to the strain itself to increase its ability to produce the antibiotic. Two types of operations can be envisaged: (1) mutation and selection and (2) genetic recombination and genetic engineering.

7.3.1.1. Mutation and Selection

The production of secondary metabolites, such as antibiotics, is controlled by more than one mechanism. The rate of production of such metabolites depends on the rates of synthesis of the metabolic products that are intermediates in the antibiotic synthesis and on the rate of their transformation. In turn, these reactions are regulated by mechanisms such as repression or induction of enzyme synthesis, allosteric regulation of the activity of the enzymes, and catabolic repression, each being determined genetically. Therefore, these mechanisms can be altered or even eliminated by mutation. Mutations occur naturally in bacterial populations during cell division but at a very low frequency. The frequency of mutation can be increased by several orders of magnitude by subjecting the strain to mutagenic agents such as ultraviolet radiation, X rays, or chemical agents (e.g., nitrous acid, nitrosoguanidine).

Since mutations are random alterations of the genetic information, they may have any kind of effect on the organism. The population of microorganisms surviving after mutagenic treatment consists of:

1. Cells in which there has been no mutation and that are therefore like the original strain
2. Cells in which one or more mutations were produced that either do not interfere with antibiotic synthesis or that inhibit antibiotic synthesis
3. Cells that have undergone mutations that increase antibiotic production

These last cells, called *high-producing mutants*, are present in the population in very low proportion. The problem consists of identifying these high producers in a mass of low producers. The process by which this is done is called *selection*.

Random Selection. This is the most empirical method of selection. Suspensions of organisms that have been subjected to mutagenic treatment are plated to give colonies derived from single cells. Colonies are selected blindly for cultivation in pure culture and then the potency (the concentration of an antibiotic produced in a fixed time of fermentation) is determined. Since mutation toward high productivity is a rare event, a large number of colonies must be tested to find high-producing mutants. A program of this type requires selection and testing of about 10,000 colonies per year. Even though this is not a very efficient or intellectually satisfying procedure, it is often the only feasible approach, at least during the early phases of development.

The efficiency of this type of selection can be increased by preselecting the potential high producers directly on the plate on which the colonies are developed. This can be done by covering the surface of the plate, after the colonies have grown, with a layer of agar containing a strain sensitive to the antibiotic. Colonies of high-producing mutants show inhibition halos larger than those around the colonies of the original strain. This method is time and cost effective but has some limitations, the main being that frequently there is no relation between the productivity of the colonies on solid medium and the productivity in liquid medium.

Selection of Morphological Mutants. Among the survivors of mutagenic treatment some colonies appear that are morphologically different from the wild type. In some cases a qualitative relationship has been found between colony morphology and antibiotic overproduction (e.g., penicillin, cycloheximide, nystatin). However, there has never been any explanation for the fact that a given morphological mutant produces more antibiotic. Since the genes governing antibiotic production are sometimes regulated by genes that also regulate aerial hypha formation and sporulation, one can surmise that alteration of these regulatory genes leads at the same time to morphological changes and increased antibiotic production.

Selection of Mutants That Are High Producers of Intermediate Metabolites. Secondary metabolites derive by transformation or polymerization of intermediate metabolites. An increased rate of synthesis of the latter might lead to greater productivity. For example, the rate of methylation of an antibiotic might be limited by the availability of the methyl donor, which is often the amino acid methionine. If the pool of methionine is increased by a mutation that leads to overproduction of this amino acid, it would lead to an increased production of the antibiotic.

Selection of Auxotrophic Mutants and Their Revertants. An *auxotroph* is a mutant that differs from the original strain in that it requires an additional nutrient. For example, an auxotroph for vitamin B_{12} must be given this substance in order to grow. It has been seen that in many cases auxotrophy has a profound effect on antibiotic synthesis. In general, auxotrophic mutants produce less than the wild strain. However, the *revertant mutants,* i.e., the strains that derive from the auxotroph and no longer require the addition of that substance to the culture medium, may sometimes be high producers. Even though we do not know the mechanism by which reversion from auxotrophy to the primary condition (*prototrophy*) causes an increase in antibiotic production, it can be hypothesized that this could be related to a change in the control mechanisms of one or more metabolic pathway, generating increased production of the antibiotic precursors and consequently a stimulation in antibiotic production.

In conclusion: (1) auxotrophy usually leads to mutants that are low producers, but in some cases auxotrophic mutants may produce more than the original strain; (2) some revertants produce more than the original strain.

Selections of Mutants with Changes in Their Metabolic Control Mechanisms. The synthesis of primary and secondary metabolites by a microorganism is regulated by many control mechanisms that act to prevent excessive production. Since the aim of improving the productive strain consists exactly in changing it so as to obtain large amounts of a given metabolite, it is easy to understand how elimination of regulatory mechanisms might be important for attaining this end. The principal control mechanisms are the following:

1. *Retroinhibition.* This consists of the inhibition of the activity (not the synthesis) of an enzyme (usually the first of a metabolic pathway) brought about by a small molecule, called the *effector*. Such an effector could be:

    a. The final product of the metabolic path, i.e., the antibiotic itself as in the case of chloramphenicol, which inhibits the activity of the enzyme arylamine synthetase. This is

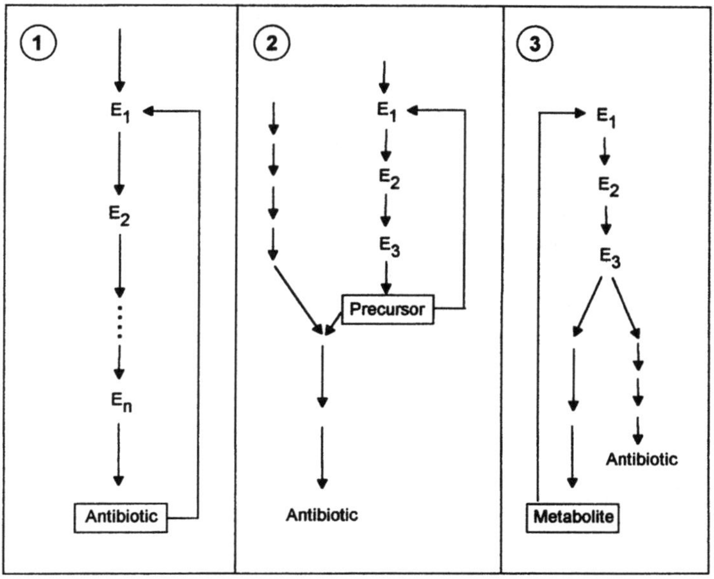

Figure 7.1. Different types of feedback inhibition.

the first enzyme of a pathway branching off the common pathway of aromatic amino acid synthesis.

b. A precursor of the antibiotic, which may feedback-regulate its own synthesis and therefore limit the rate of synthesis of the antibiotic.

c. A metabolite sharing a part of the antibiotic-synthesizing pathway (see Figure 7.1).

2. *Repression of synthesis of one or more enzymes*. This is the inhibition of the rate of synthesis of one or more enzymes specifically caused by a small molecule, the *repressor*, which usually is the final product of the synthetic pathway of the antibiotic, or a molecule structurally similar to this product.
3. *Induction of the synthesis of one or more enzymes*. This is an increase in the rate of synthesis of one or more enzymes caused specifically by a small molecule that is generally the substrate of the enzyme or a structurally similar molecule. A number of molecules has been identified, called *autoregulators*, whose presence at very low concentrations induces antibiotic production, directly or indirectly. The best known and studied of these regulators is

factor A, initially isolated from *Streptomyces griseus*, producer of streptomycin.

4. *Glucose-catabolite repression*. This is a decrease in the rate of synthesis of one or more enzymes, especially those of degradative metabolism in the presence of glucose or other sources of easily utilizable carbon. The synthesis of many antibiotics is often controlled by this catabolic repression, but the biochemical mechanism of repression in primary metabolism seems different from that of repression in secondary metabolism.
5. *Nitrogen-catabolite repression*. This is a decrease in the rate of synthesis of enzymes related to the catabolism of nitrogenous compounds (e.g., proteases, amidases, nitrate reductase) in the presence of easily utilizable nitrogen sources, such as ammonia. The synthesis of many antibiotics is controlled by this type of mechanism.
6. *Repression by inorganic phosphate*. Often the addition of inorganic phosphate to a culture actively synthesizing an antibiotic strongly decreases the rate of synthesis. Several mechanisms have been postulated to explain the phosphate regulation of antibiotic biosynthesis:

    a. Phosphate stimulates primary metabolism, thus channeling energy and substrates toward growth instead of synthesis of secondary metabolites. In this respect it is known that phosphate shifts carbohydrate catabolism from the hexose monophosphate pathway to the glycolytic pathway.
    b. Phosphate controls the synthesis of a specific inducer of the antibiotic synthesis.
    c. Phosphate may inhibit the activity or repress the synthesis of phosphatases involved in the metabolism of phosphorylated precursor(s) of the antibiotic.

It is conceivable that different mechanisms may be operative in different organisms or under different physiological conditions and that more than one might be operative simultaneously. The alteration of one or more of the many control mechanisms is likely to be one of the most productive types of mutation, especially in the early stages of strain development.

Selection of Strains Resistant to the Antibiotic. Producing strains are often sensitive to some concentration of their own antibiotic product.

In some cases the mutants that are resistant to the toxic effects of the antibiotic are also high producers.

Selection of Strains Resistant to Antibiotic Precursors. The classic case is the production of penicillin V after addition of phenylacetic acid to culture broth. The precursor is toxic for the producing strain. The strains resistant to high concentrations of phenylacetic acid are often high producers of penicillin V.

7.3.1.2. Genetic Recombination and Genetic Engineering

The rate of production of an antibiotic is determined by the slowest of the reactions of its biosynthetic pathway. When there are two producer strains sufficiently different in their genetic structure, it is likely that the limiting reaction is not the same in the two organisms.

After a genetic recombination of these strains, in some recombinants the limiting elements may be substituted by the more efficient corresponding elements, resulting in a higher rate of production, which might, in turn, be limited by a third biosynthetic step.

Several techniques have been developed to achieve recombinations between antibiotic-producing microorganisms. In many cases conjugation among different strains of *Streptomyces* has been obtained. However, no important results for productivity have as yet been reported.

The technique of protoplast fusion has yielded more fruitful results. In fact, in streptomycetes it is possible to obtain the formation of protoplasts by treating the cells with lysozyme under suitable conditions. Protoplasts of different strains can be fused in the presence of polyethylene glycol and then, on a suitable culture medium, induced to regenerate viable cells. During fusion an extensive recombination of genomes occurs and the regenerated cells contain a large variety of different genotypes. With this method, a marked increase of productivity has been obtained in many cases.

Genetic engineering techniques are used, currently, to obtain higher-producing strains. Several genes responsible for the biosynthesis of various antibiotics have been cloned and it has been found that most often regulatory genes are adjacent to the structural ones. One can then envisage the possibility to selectively alter these regulatory genes or duplicate them in the case of a positive regulation by applying techniques of site-directed mutation. Alternatively, one can substitute the promoters of the biosynthetic genes with more efficient promoters, obtaining in this way the desired increase of the genic product.

Although these concepts are still in the exploratory and academic stage, in many laboratories experiments are being carried out aimed at the practical application of increasing production yields of commercially important antibiotics.

7.3.2. Improvement of Fermentation Conditions

The rate of production of an antibiotic is also dependent on many nutritional factors: the nature and concentrations of the carbon and nitrogen sources, the concentration of inorganic salts (phosphate, sulfate, etc.), the concentrations of various substrates, the pH of the medium, the partial pressure of oxygen and carbon dioxide dissolved in it. These factors can affect antibiotic production *directly* by interference with metabolic control mechanisms in their roles as precursors or effectors in biosynthetic pathways, or *indirectly* by regulating the rate of cell growth. By manipulating systematically the components of the culture medium and the physical conditions of fermentation (e.g., aeration, agitation, temperature) it is possible to increase productivity, i.e., the production of antibiotic per unit of microbial mass per unit of time.

7.3.2.1. The Scaling-Up Problem

The improvement of fermentation conditions is studied with laboratory instrumentation and equipment, where producing organisms are grown in vessels containing from a few tens of milliliters to a few liters. Industrial production, to be economically feasible, must take place in fermenters containing up to tens of cubic meters. There is then the problem of converting the optimal fermentation conditions from the laboratory scale to the industrial scale. This process is called *scaling-up* and is concerned with two different correlated aspects: (1) designing industrial fermenters that can reproduce laboratory conditions and (2) modifications of the laboratory fermentation process to adapt it to the different, large-scale hardware.

The need for and the difficulties involved in this last point derive from the fact that one cannot always increase the various parameters proportionally. For example, aeration and the energy needed for stirring vary almost proportionally with the volume of the fermenter, but they depend also on the shape of impellers. Heat dispersion is proportional to surface area, whose ratio to fermentation volume varies with fermenter size. The varying heights of the liquid layers in fermenters of diverse size affect the time and the extent of carbon dioxide exchange. The development of a microorganism from the initial colony to a broth containing an equal number of cells per cubic centimeter requires, for the industrial plant in respect to the laboratory scale, not only more time, but a larger number of generations, and this may cause differences in productivity, because of the accumulation of spontaneous mutants.

These problems are first attacked in the so-called *pilot plant*. This consists of a series of fermenters of varying capacities, ranging from some tens of liters to some cubic meters. In these vessels fermentation conditions approximating the industrial ones can be studied. The fermenters are usually equipped with sensors that allow the various fermentation parameters to be measured and recorded, such as glucose and nitrogen levels, pH, and dissolved oxygen concentration. Some of these parameters can be modified during the fermentation by control systems, manually or automatically.

In the last few years, a branch of bioengineering has been developed concerned with identifying the parameters important in the extrapolation from laboratory experiments to an industrial scale.

Chapter 8

The Use of Antibiotics

Antibiotics are among the most important and most frequently used drugs. In this chapter, the principles and the basic concepts of chemotherapy, i.e., the use of antibiotics to treat or prevent infectious diseases, are delineated. The reader interested in the specific use of single antibiotics and treatments of the various infectious diseases is directed to the vast specialized literature.

In addition to therapeutic use, antibiotics have found important applications in different fields, such as *animal husbandry* and *agriculture,* and are essential tools for research in *biochemistry* and *genetics.* These aspects are also briefly discussed in this chapter.

8.1. Chemotherapy of Infectious Diseases

The expression *chemotherapy of infectious diseases* refers to the treatment of diseases of microbial origin by the systemic administration of antibiotics or other antimicrobial agents. Basically, this aims at inhibiting the multiplication of the infecting microorganism through *selective tox-*

icity, without interfering with the function of the host. The host's defense mechanisms are thereby able to overcome the infection. It is beyond the scope of this book to undertake a detailed analysis of the methods or of the uses of chemotherapy, i.e., which particular antibiotic should be used for a given infection and how it should be administered. Here we present only the basic principles of chemotherapy, which can be stated briefly in the following way: *An antibiotic is therapeutically effective, when used to treat or prevent an infection, if (1) the infection is caused by a susceptible microorganism, (2) it can reach sufficient concentrations in tissues to inhibit growth of the infectious agent, (3) the treatment is continued for a sufficient time, and (4) it does not cause severe adverse reactions.*

8.1.1. Principles of Chemotherapy

The various aspects to consider in selection of an antibiotic are outlined in Table 8.1.

8.1.1.1. Microbiological Aspects

Antibiotics should be used only when the disease to be treated is infectious (either demonstrated or on the basis of clinical evidence). No single antibiotic is active against all bacterial species, and, in fact, each antibiotic has its own specific spectrum of activity. In addition, within a single species of bacteria some strains vary in their susceptibility to the same antibacterial agent, and some bacterial strains that are initially

Table 8.1. Aspects to Consider When Deciding on an Antibiotic Treatment

Aspect	Considerations
Microbiological	Nature of infecting agent Its spectrum of sensitivity to antibiotics Dosage, interval between doses, duration of treatment
General clinical and toxicological	General conditions of the patient: age, pregnancy, genetic factors, concomitant illnesses and treatments, liver and kidney functions, conditions of immune system, etc.
Adverse reactions	Superinfections, avitaminosis, immune reactions, local intolerance, etc.
Epidemiological	Frequency of single and multiple resistants in environment in which patient lives Prevention of selection of resistant mutants and of spread of resistance

susceptible to a given antibiotic can become resistant through one or more of the mechanisms described in Chapter 4. For these reasons, under ideal conditions, before beginning treatment with an antibiotic, one should know which organism is responsible for the infection and what its spectrum of susceptibility is to different antibiotics, i.e., its antibiogram (see Chapter 2).

To obtain an antibiogram requires time and the availability of a specialized laboratory. When it is deemed necessary to begin treatment before the results of the antibiogram are available, the antibiotic is chosen on the basis of the clinical characteristics of the infection, and may then be changed as soon as the results of the susceptibility determinations are available.

In some cases the clinical characteristics of the infection strongly suggest what the pathogenic agent may be, and experience indicates the *antibiotic of choice* against that microorganism. For example, typhoid fever is caused by *Salmonella* and treated with chloramphenicol, and scarlet fever is caused by streptococci and is normally treated with penicillin G or V. In other situations the choice is more complicated because the same type of illness may be caused by several infectious agents (e.g., septicemia and bacterial pneumonia) or by microorganisms that are susceptible to several antibiotics, so that the choice must be based on epidemiological, clinical, and pharmacological considerations. Ideally, the antibiotics used should inhibit growth of the infecting agent without interfering with the nonpathogenic bacterial flora, to avoid secondary effects. In other words, the antibiotic should be specific for the etiologic agent causing the infection. This is in contrast with the current tendency to prefer broad-spectrum antibiotics over those with narrower spectra. The preference for broad-spectrum antibiotics is well founded only when one suspects or has shown that more than one bacterial species with different spectra of sensitivity (see also Section 8.1.3) are or could be involved in the infection.

Generally, it is not important whether the antibiotic is bactericidal or bacteriostatic, since its function is to block the spread of bacteria and permit the body's defense mechanisms (cellular and immunological) to rid the body completely of the infectious agent. However, when the immune defenses are not very effective or in chronic conditions, a bactericidal antibiotic may be needed.

8.1.1.2. Toxicological Aspects and Secondary Effects

Although the microbiological aspects of the infection are of paramount importance, they are not the only parameters the physician must take into consideration in choosing an antibiotic. There are numerous other aspects to be considered but they are beyond the scope of this book. In this chapter we will discuss the possibility of adverse reactions that can be caused by antibiotic treatment.

We have pointed out many times that antibiotics are a widely diverse group of chemical substances, so that there is no toxic or adverse reaction that may be considered typical for the entire group, and the contraindications for each family of antibiotics must be considered separately. The only secondary effects that may be considered to be group-specific are those that result from changes in the normal symbiotic bacterial flora. Broad-spectrum antibiotics, especially when given orally, may cause a massive alteration of the intestinal bacterial flora with the possible consequence of an abnormal proliferation of insensitive organisms, either fungi or resistant bacteria, giving rise to *superinfections*. Usually, these are not very severe, but in some instances they require suitable medical treatment. It should be noted that the same effect on the intestinal flora can occur when a broad-spectrum antibiotic is given intramuscularly or intravenously if the antibiotic is excreted into the bile in an active form.

A less important negative effect caused by the alteration of intestinal flora is vitamin deficiency, which is sometimes seen during prolonged treatment and which is the result of elimination of bacteria that synthesize vitamins. This inconvenience is easily overcome by vitamin administration.

These phenomena are not limited to the intestinal tract (there may be, e.g., a superinfection of the vagina by *Trichomonas* resulting from destruction of the resident saprophytic flora) and they are rarely seen when narrow-spectrum antibiotics are given, because these affect only a fraction of the bacterial flora.

The majority of antibiotics have very low toxicity. The cellular and molecular bases for this characteristic, which are clearly related to the selectivity of action, were discussed in Chapter 3.

However, some families of antibiotics do have toxic effects or special side effects, such as the ototoxicity of aminoglycoside antibiotics or the photosensitization of tetracyclines, and it is important to take these into account in clinical practice as possible contraindications. In addition, as any drug or any foreign substance taken into the body may do, antibiotics can stimulate immune phenomena, with production of antibodies and hypersensitivity or allergic reactions. The frequency and the severity of these reactions vary greatly from one family of antibiotics to another and can also be quite different for different members of the same family. Hypersensitivity and allergic reactions may take different forms. The less severe are usually skin manifestations such as rashes, which usually do not necessitate discontinuing the treatment. Fever and leukopenia are more serious. The most serious are aplastic anemia, angioedema, and anaphylactic shock. An antibiotic or antibiotics belonging to the same family should never be given to a patient who has already demonstrated hypersensitivity to it, but this is not a contraindication for giving antibiotics of a different family.

8.1.1.3. Pharmacokinetic Aspects

Antibiotics are effective only if they are present at the infection site at inhibitory concentrations. The dose and the frequency of treatment are recommended on the basis of careful pharmacological and clinical experimentations, to guarantee that this condition is fulfilled. Too low doses or adequate doses at too long intervals may not achieve inhibitory concentrations at the site of infection, and thus they will be ineffective. In addition, insufficient doses of antibiotic or irregular treatments may produce in the body a situation similar to that used in the laboratory to select multistep resistant mutants (see Chapter 4).

8.1.1.4. Epidemio-Ecological Aspects: Resistance

In many countries, *Streptococcus pneumoniae* and *Treponema pallidum* have not yet, even after 50 years of therapeutic use, developed any appreciable degree of resistance to penicillin (strains of *S. pneumoniae* resistant to penicillin have been seen in South Africa, Spain, and, more recently, in the United States). In contrast, some streptococci and most of the staphylococci have become resistant to various antibiotics to which they were originally susceptible. This has occurred also for several pathogenic gram-negative bacteria.

Therefore, the physician must keep up with the latest information about the distribution of resistance of different bacteria to different antibiotics in the geographical area in which he or she is working. Thus, a physician will not give penicillin for treatment of a staphylococcal infection if the majority of the staphylococcal population in the region is resistant to this antibiotic, unless the strain isolated from the patient is first demonstrated to be penicillin-susceptible.

In this respect, it should be noted that:

1. The distribution of resistance to an antibiotic in a given bacterial population does not remain constant with time but varies with the antibiotic use in that geographical region.
2. The frequency of resistant strains increases with the selective pressure, i.e., when the use of the antibiotic increases.

3. The frequency of strains with transferable resistance is a function of the concentration of the microbial population.

For these reasons, the highest frequency of strains resistant to one or more antibiotics (see Chapter 4, on multiple resistance) is found in hospitals where there is the highest use of drugs and the highest concentration of pathogenic microorganisms.

8.1.2. Prophylaxis

Antibiotics are often used to prevent an infection from occurring clinically. Although there is little doubt that preventing a disease is better than curing it, we must not forget that the indiscriminate use of antibiotics as prophylactic agents may cause more damage than benefit, by selecting resistant strains, causing adverse effects, and sensitizing patients. Rational use of antibiotics in prophylaxis requires:

1. A condition in which infection is expected to occur. Examples are:
    a. Extensive burn trauma (infections by *Staphylococcus aureus* and *Pseudomonas aeruginosa*)
    b. Certain surgical operations such as open-heart surgery or hip surgery
    c. Streptococcal infections of the throat that may result in rheumatic fever
    d. Healthy carriers of meningococci and *H. influenzae**
2. Correct choice of the antibiotic to be administered. Since a well-defined infection is expected, it is not difficult to select the appropriate antibiotic on the basis of the sensitivity of the causative microorganism.
3. Correct timing of administration. The ideal timing is that which ensures the presence of the antibiotic at the site of the possible infection when this may be initiated. This time can easily be predicted for surgery.

In other cases, such as rheumatic fever, because of the impossibility of predicting the time of reinfection, antibiotic coverage must be extended over a long period of time.

8.1.3. Combinations of Antibiotics

The expression *antibiotic combination* is used to indicate the simultaneous administration of two or more antibiotics. The practice of com-

*This is not an exhaustive list and other examples could be added.

bining several antibiotics or of combining antibiotics with other chemotherapeutic agents is as old as antibiotics themselves. Many reasons have been proposed (microbiological, clinical, or the simplistic idea that two drugs are better than one) for the use of many different combinations in place of single antibiotic treatments, but not all are acceptable.

A rational combination of two or more antibiotics may have the following aims.

8.1.3.1. To Avoid Selection of Resistant Bacteria

The frequency of mutation toward resistance to two antibiotics that do not present cross-resistance is the product of the frequency of resistance to each antibiotic. Thus, if the frequency of resistance to antibiotic A is, for example, 10^{-7} and that to antibiotic B is 10^{-8}, the frequency of appearance of resistance to both A and B at the same time is $10^{-7} \times 10^{-8} = 10^{-15}$. In the combined treatment with A and B, the organism resistant to A will be eliminated by B and the organism resistant to B by A. Only the organisms resistant to both A and B will not be inhibited by the combination, but these occur with an extremely low frequency, negligible for all practical purposes.

Rational use of combinations having this aim requires the infecting strain to be sensitive to both antibiotics, and both antimicrobial agents to be present at the site of infection at the same time. Consequently, their pharmacokinetics must be well studied and the dosages must be carefully planned.

The combined treatment with two or more antibiotics, following these principles, has yielded excellent results in the therapy of tuberculosis.

8.1.3.2. To Enlarge Antibacterial Spectrum of the Treatment

The reasoning here is more or less the following: if one combines two antibiotics with limited but different spectra of antibacterial activity, the sum of the two spectra will be a broader spectrum. These types of combinations are normally suggested for mixed infections, such as otitis media, peritonitis, etc. The antibiotics must be chosen on the basis of their spectra of action and must not be antagonistic.

8.1.3.3. To Increase Therapeutic Effectiveness

When a homogeneous bacterial population is treated simultaneously with two antibiotics, there may be three types of results: synergism, additivity, and antagonism (see Chapter 2). With synergism, the antibiotic effects of the combination are greater than the sum of the effects of each antibiotic alone. Additivity refers to an effect of the combination that is equal to the sum of the effects of the two. In the case of antagonism, one of the components decreases the effectiveness of the other. Although these three situations are easy to distinguish conceptually as separate entities, it is often complicated to measure them quantitatively. In addition, interference of one antibiotic with another may or may not be seen, depending on the parameter measured. For example, a product may negatively influence the bactericidal action of another without changing the MIC. Most importantly, the results obtained *in vitro* are not always confirmed by the experimental infections or by clinical data.

In case of synergism the concentration of each antibiotic in the combination needed to obtain inhibition of bacterial growth is only a fraction of the MIC for each antibiotic when used alone. Therefore, in a combination it is possible to treat an infection with lower doses of each antibiotic.

The advantages can be summarized as follows: (1) possibility of using antibiotics with high MICs, not practically attainable at the site of infection; (2) reduction of the dose of antibiotics having a certain degree of toxicity. It must be said that the latter consideration is valid only for the antibiotics whose toxicity is dose-dependent.

In any case, it should be noted that combinations can be either more toxic or toxic in a different way than the components given singly. In addition, the presence of one drug can affect the pharmacokinetics of the other (absorption, distribution, metabolism, and excretion). Therefore, every new combination must be considered to be a new drug and must be submitted to animal toxicology studies and to studies of its pharmacokinetics and metabolism in man.

8.2. Uses Other Than Human Pharmacology

In addition to their uses in humans, antibiotics are widely used in animal husbandry, in veterinary medicine, and, in some countries, in agriculture. In these areas, antibiotics are also used sometimes for purposes that are not strictly therapeutic and according to principles different from those valid for human medicine. In this section we briefly describe these different uses with their advantages, disadvantages, and potential risks. As in the preceding sections, we do not give a detailed description of the uses of individual antibiotics in veterinary practice and agriculture, but the discussion is limited to general usage principles.

8.2.1. Uses in Veterinary and Animal Husbandry

8.2.1.1. Prophylaxis and Therapy

In veterinary medicine antibiotics are widely used in prophylaxis and treatment of diseases as part of programs designed either for individual or for mass treatment of animals.

Individual Treatment. This type of intervention is practiced in treatment of small (e.g. dogs, cats, other pets) and large (e.g., cows, sheep) animals suffering from specific infections (e.g., pulmonitis, bronchitis, mastitis). In the treatment of small animals, the same criteria applied to chemotherapy of infections in humans are applicable and the same antibiotics are used. In the treatment of large animals, the principles of human chemotherapy are widely applicable; however, for animals designed for food production (meat, milk, and eggs) it is necessary to withdraw the antibiotic for a sufficient period of time before slaughter or before egg collection, to avoid ingestion of the antibiotics along with the food by the population.

Mass Treatment. This type of treatment, prophylactic or therapeutic, is given to some or all of the animals present in a facility that has been affected by an infection. To farm animals (chickens, pigs, calves) the antibiotics are given in the feed or the drinking water (medicated supplements).

Mass treatment has the following advantages over individual treatment: (1) simultaneous treatment of all animals interrupts the infectious cycle and eliminates the foci of infection; (2) treatment of animals with active infections or with asymptomatic or presymptomatic infections makes it possible to achieve cures without aftereffects that may reduce the economic value of the animals; (3) treatment is facilitated and controlled as the antibiotics are given in the feed or drinking water.

Mass treatment has the following disadvantages: (1) it is difficult to obtain exact individual dosage and (2) there are difficulties related to the types of feed, the cycles, and the technology of breeding.

Even with these limitations, mass treatment is at present considered the most suitable and economic system for an effective attack on infectious diseases in intensive animal-raising facilities.

8.2.1.2. Improvement of Production in Animal Husbandry

Antibiotics have been widely used for several years in animal nutrition to increase the growth rate of the animals (auxinic use). As this has no parallel in human medicine it is necessarily discussed separately.

In 1948 it was first noticed that incorporation of small doses of antibiotics in the diet increased the weight gain in chickens. These experiments were confirmed innumerable times with a variety of antibiotics and different species of animals. As we have stated in many instances, the various antibiotics and chemotherapeutic agents have nothing in common except their ability to inhibit bacterial growth, so the effect of antibiotics on weight gain in animals must be mediated through the intestinal bacterial flora. This hypothesis was confirmed by the discovery that the growth rate of germfree animals raised under sterile conditions does not show any response to antibiotic treatment. When germfree animals are artificially infected with bacteria present in the intestine of normal animals, their growth rate can be stimulated by incorporation of antibiotics in the diet. This effect on growth is probably related to one or more of the following causes:

1. Inhibition of intestinal bacteria that produce toxins
2. Inhibition of bacteria that cause asymptomatic diseases
3. Inhibition of bacteria that destroy or sequester proteins or other essential nutrients in the diet
4. Stimulation, as a consequence of inhibition of part of the flora, of those bacteria that synthesize nutritional factors needed for growth of the host

In the last analysis, the effect of the antibiotic consists of modifying the system constituted by the intestinal flora, with consequent alteration in the physiology of the animal. The final result is a greater rate of weight gain and a better conversion factor (less food consumed for an equivalent weight increase), which is translated into consistent economic savings.

8.2.2. Agricultural Use

A certain number of antibiotics have been developed for agricultural use. The more important applications can be summarized as follows:

1. Control of bacterial infections. To this purpose, antibiotics generally developed for human medicine are used, in particular, streptomycin against infections by *Erwinia*, *Xanthomonas*, etc.
2. Control of Fungal Infections. "Ad hoc" antibiotics have been developed for these infections, which are widely diffused and very important economically. Many of these belong to the nucleosides, such as blasticidin S, active against *Piricularia orizae* (rice pathogen), and the polyoxins. Other important antifungal antibi-

otics are cycloheximide, a known inhibitor of protein synthesis in eukaryotes, which, however, has a certain toxicity for plants, and kasugamycin, an aminoglycoside.

3. Herbicides. Among the numerous synthetic substances that show antibiotic properties, some are used for the control of infestant plants. The most studied of these is bialaphos, an inhibitor of glutamine synthetase.

8.3. Antibiotics as Research Tools

Many antibiotics, both those used in practice and those that have no therapeutic value, have been and continue to be used as biochemical tools. There is an enormous amount of literature on the subject, but we will mention only a few significant examples. The usefulness of antibiotics in this respect lies mainly in their specific mechanisms of action, as illustrated by the following applications.

1. Their ability to inhibit a definite informational molecule or enzyme makes them very valuable for understanding this molecule's functions in cellular metabolism. Many details in protein synthesis have been elucidated by using such inhibitors as cycloheximide and chloramphenicol. The first indication of a difference between the RNA polymerases of eukaryotes and prokaryotes was suggested by their different sensitivity to rifamycins. The degree of autonomy of mitochondria and chloroplasts was made clear when sensitivity to specific antibiotics was observed.
2. A complex task that geneticists started several years ago and that is still under way is to draw the genetic map of microorganisms, i.e., to establish the relative locations of genes on their DNA. Again antibiotics have proved to be invaluable tools. In fact, in all cases in which the resistance to an antibiotic is the result of a modification of the target enzyme, locating the position of the "resistance character" in the genome is equivalent to locating the position of the enzyme. For example, the position of the RNA polymerase genes on the *E. coli* chromosome was established by mapping the character "resistance" to rifampin in this microorganism.
3. Several genetic operations make use of antibiotics, either as genetic markers (a definite pattern of resistance to some antibiotics makes a strain easily identifiable) or for selecting a microorganism with given characteristics, e.g., in conjugation experi-

ments. Moreover, penicillin is used in selection methods where the microorganisms able to grow under given conditions must be eliminated. Nongrowing cells can be positively selected in this way.

4. In all procedures of genetic engineering, the vectors used for transferring genes from one organism to another contain determinants for resistance to one or more antibiotics in order to facilitate the selection of the recombinant clones.

Chapter 9

Antibiotics and Producer Organisms

The relationship between antibiotics and their producer microorganisms is a matter of lively discussion among specialists. Why is the ability to produce a great variety of antibiotics limited to relatively few classes of microorganisms? How can a producer organism survive in the presence of a substance toxic to it? What is the "natural" function of antibiotics, i.e., what evolutionary advantage do they confer to the producer? There are no generally accepted answers to these questions.

Some considerations and hypotheses are presented in this chapter, with preference for those that we subjectively consider most convincing.

What is the "natural" function of antibiotics? What evolutionary advantage does a microbial strain gain by its ability to produce an antibiotic substance? Why is the ability to produce antibiotics more common in certain taxonomic groups than in others? During more than forty years of intense antibiotic research, several hypotheses have been put forth to answer the above questions. In evaluating such hypotheses, it is important to remember that antibiotics are a heterogeneous class of compounds with respect to chemical structure, biosynthetic origin, and mechanism of action, and that they are the products of quite different organisms. Therefore, it is conceivable that their function depends on the type of antimicrobially active metabolite, and that different metabolites may confer quite diverse evolutionary advantages on the producing organisms.

Before analyzing the various hypotheses, a discussion on the accepted relationships between the taxonomic positions of the producing organisms and the nature of their antibiotics is necessary.

9.1. Classes of Antibiotics and Taxonomic Position of Producing Organisms

It would be interesting to be able to establish a relationship between the taxonomic position of a producing strain and some properties of the antibiotic it produces, such as chemical nature, antimicrobial spectrum, or mechanism of action. Several analyses have been made with this aim in mind. However, as more antibiotics were discovered and as their characteristics became better defined, it became clear that no simple relationship exists. However, some general rules have been defined as follows:

1. Most antibiotics are products of the secondary metabolism of three main groups of microorganisms: eubacteria, actinomycetes, and filamentous fungi. Only a few antibiotics are produced by higher fungi, algae, and plants, and they generally show low antimicrobial activity and little specificity.
2. The actinomycetes produce the largest number and the greatest variety of known antibiotics (more than 6000 substances have been isolated from them). The lower fungi produce several kinds of secondary metabolites, of which approximately 1500 show antibiotic activity. The eubacteria (mainly spore-forming bacilli and pseudomonads) produce a fair number of antibiotics (about 1000 have been described to date). Most of those produced by bacilli belong to one class, the polypeptides.

   We may add that another group, the myxobacteria, although little studied, has revealed a high frequency of production.
3. Frequently, a given strain produces a "family" of structurally and biosynthetically related substances (see Chapter 6). In addition, a given strain may produce two or more unrelated antibiotics.

4. Antibiotic production is not rigorously species-specific. In fact, different strains of microorganisms belonging to the same species may produce completely different antibiotics. A classic example is *Streptomyces griseus*. Streptomycin (an aminoglycoside), novobiocin (a glycoside with a complex aromatic moiety), cycloheximide (aromatic structure derived from acetate units), viridogrisein (a depsipeptide), griseoviridin (a lactone), candicidin (a polyene), and grisein (a sideromycin) are produced by different strains of this species. The same antibiotic molecule can be produced by strains belonging to different taxonomic groups. For example, cycloserine has been isolated from both a *Streptomyces* and a *Pseudomonas* strain and penicillin N is produced by lower fungi as well as by actinomycetes.
5. The greater the taxonomic difference between two microorganisms, the lower the probability that they produce the same antibiotic molecule.
6. In addition to the above rules, a relationship appears to exist between the taxonomic group of the producing organism and the biosynthetic pathways of the antibiotics produced. Nevertheless, further research is needed to clarify this relationship. Some biosynthetic pathways of secondary metabolism occur generally (e.g., the ability to activate and to condense amino acids to produce polypeptide antibiotics is found in eubacteria, actinomycetes, and lower fungi); others are present in only one of the three groups (e.g., virtually all the known secondary metabolites originating from terpene synthesis are produced by fungi). Even within actinomycetes there seems to be biosynthetic differentiation: for example, the biosynthesis of aminocyclitols, and therefore of aminocyclitol-containing antibiotics (aminoglycosides), is found much more frequently in the genera *Streptomyces* and *Micromonospora* than in other genera of the order Actinomycetales, such as *Nocardia* and *Actinoplanes*. However, it should be noted that this relationship has only a statistical and not an absolute value. We surmise that, with increased attention and expanded research on metabolism of antibiotic-producing organisms, it will become possible to establish a relationship between the biosynthesis of secondary metabolites (including antibiotics) and unique features of the producer organism's metabolism. For example, the ability (typical of fungi) to produce terpene antibiotics might be correlated with the presence of the metabolic pathway (condensation of isoprenic units) utilized by these organisms to produce the sterols of their cell membranes.

9.2. Paradox: How to Avoid Suicide

Most antibiotics are active against the organisms that produce them. In fact, no growth is observed when a suspension of spores or mycelium fragments of a producer organism is inoculated into a fresh medium containing the antibiotic that the organism normally produces. We are dealing here with the apparent paradox of the ability of a strain to produce a large quantity of a substance that at low concentration inhibits its growth. This paradox is solved by one or more of the following mechanisms:

1. Repression of antibiotic synthesis during growth
2. Alteration of permeability
3. Inactivation of the antibiotic
4. Alteration of the intracellular target of the antibiotic

9.2.1. Repression of Antibiotic Synthesis

Frequently, antibiotics are produced only after completion of the growth phase. This observation has led to the terminology *trophophase* and *idiophase*. The former refers to the period of vegetative mycelial growth, the latter to the period of antibiotic synthesis. Idiophase is often repressed by substances that favor rapid cellular growth, such as glucose (catabolite repression), ammonium ions (nitrogen repression), and phosphate. When the levels of these nutrients are low, the rate of cell growth is slowed and antibiotic synthesis is derepressed. In this way, the antibiotic is synthesized only when the producing organism is physiologically insensitive to it.

9.2.2. Alteration of Permeability

It has been shown that certain high-producing strains differ from low-producing strains in their permeability to the antibiotic they produce, i.e., they are able to efficiently excrete the antibiotic and, when growing, they lose the ability to assimilate it from the medium. In the case of tetracyclines, a membrane protein has been identified that efficiently carries the antibiotic out of the cell.

9.2.3. Antibiotic Inactivation

Even in the presence of the above-mentioned mechanisms the cellular concentrations of the antibiotic may reach inhibitory levels, and in this case suicide can be avoided by enzymatic inactivation of the antibiotic. The two more common mechanisms of inactivation are acetylation of an amine and phosphorylation of a hydroxyl group. The former case has been observed in the production of puromycin, the latter one has been particularly studied for the producer of streptomycin. Both mechanisms are operative in neomycin biosynthesis and have been observed for other aminoglycosides. It is noteworthy that the inactivation is temporary as the antibiotic is converted into an active form during excretion.

A different form of inactivation is shown by the production of a specific protein that binds to the antibiotic with high affinity. For example, the strain producing bleomycin expresses a small acidic protein that binds to the antibiotic, thus preventing its interaction with its target, DNA.

9.2.4. Alteration of the Target of Antibiotic Action

In a number of cases the resistance of the microorganism to the antibiotic produced is the result of an alteration of the macromolecule that is the antibiotic's target. This alteration can be constitutive or induced by the presence of the antibiotic. The most studied cases are the alterations of the ribosomal RNA in strains producing antibiotics that inhibit ribosomal functions. The most frequent alteration is methylation of rRNA at specific nucleotides. In the cases of macrolides and lincomycin, and gentamicin and kanamycin, the methylated bases were identified with certainty as localized in the 23 and 16 S rRNAs, respectively.

Another example of target alteration, found in the strain producing novobiocin, is the expression, simultaneously with antibiotic production, of a second gyrase (in addition to that expressed during growth), which is insensitive to the antibiotic.

9.2.5. Combined Mechanisms

Frequently, more than one mechanism is operative in the same strain, as is the case for the high-producing mutants of rifamycin B. Rifamycin S, the central metabolite of rifamycin biosynthesis, which is a toxic compound, is inactivated by condensation with a glycol unit to produce rifamycin B, a nontoxic compound. Rifamycin B seems to be excreted more efficiently than rifamycin S. In addition, the RNA polymerase (the target of rifamycins) of high producers is less sensitive to rifamycin S than is the enzyme of the wild-type strain.

Another case of combined mechanisms can be considered the inactivation of the antibiotic during biosynthesis accompanied by an alteration of the permeability, which prevents the reuptake of the antibiotic after it has been excreted in the environment.

9.3. Hypothesis about the Functions of Antibiotics in Producing Organisms

It was stated in Chapter 6 that antibiotics are secondary metabolites and that, as such, do not have any observable function in the development of the producer microorganism. So the obvious question is: what, if any, is the evolutionary function of these products? This question has raised different answers and lively discussions. The different points of view can be classified into three main groups:

1. Metabolic functions independent of the antimicrobial activity
2. Regulation functions of the producer strain metabolism
3. Inhibition of other microorganisms

Before discussing these points, some general considerations must be put forward; first, as mentioned in Chapter 6, antibiotics are part of the larger class of secondary metabolites and, as not all the secondary metabolites possess antibiotic activity, the function of antibiotics in nature can be different from that of the inactive secondary metabolites. Second, as already discussed, antibiotics are not only different in their spectra of activity and in their mechanisms of action but also are produced by different genera of microorganisms. And third, although the two aspects may be connected, one should distinguish the function that antibiotics may have (i.e., the evolutionary advantage they confer to the producer) from their origin (i.e., the evolutionary mechanism leading to their production).

9.3.1. Metabolic Functions Independent of Activity

Many authors have suggested that the activity of antibiotics may be adventitious and that it does not confer any specific advantage to the producer microorganism. The production of these substances would be related to the necessity to eliminate excess metabolites that would accumulate during certain phases of metabolism, either because of a nutritional stress or because of a mistake in metabolism regulation. These hypotheses are not easily tenable for the following reasons:

1. The number of secondary metabolites possessing a specific biological activity is very high.

2. The quantity of antibiotic produced by wild-type microorganisms is generally very small when compared to the pools of primary metabolites. In addition, since biosynthetic precursors are intermediate metabolites such as acetic acid or amino acids, there is no definite need to transform these in complex and useless molecules instead of polymerizing them as storage products.
3. The organization of the biosynthetic genes, as shown in all known cases, suggests the existence of a total functionality aimed at the production of specific substances.

9.3.2. Regulation of the Producer Strain Metabolism

According to this hypothesis, antibiotics would be regulators of some metabolic functions. There is a time relationship, verified in many cases, between the beginning of antibiotic production and sporulation. Therefore, it was hypothesized that the function of antibiotics is to block some metabolic pathways to induce differentiation. This hypothesis is supported by the observation that in many cases nonproducing mutants are also unable to sporulate and to form aerial mycelium.

However, the body of information now available weighs against this hypothesis. In fact:

1. Many antibiotics have no influence on the metabolism of the producing organisms (e.g., penicillins, which are produced by fungi, are active only against bacteria, and polyenes, which are produced by streptomycetes, are active only against eukaryotes).
2. As discussed in the previous section, if the antibiotic is active against its producer there are mechanisms that protect the producer from the antibiotic action. These mechanisms are activated at the same time as antibiotic biosynthesis.
3. Studies on the genetics of sporulation and of antibiotic production have demonstrated that in many cases both of these processes can be considered parts of the larger differentiation process. They are both initiated by activation of the same regulatory genes but then proceed independently under the control of different genes. In fact, it is often possible to obtain nonproducing mutants able to sporulate or mutants blocked in various stages of differentiation able to produce antibiotics.
4. In spite of the numerous studies and experiments carried out for this purpose, it has never been demonstrated with certainty that antibiotics are able to induce or favor differentiation. This is in contrast with what is observed with other secondary metabolites, such as Factor A in *Streptomyces griseus* and 6-methylsalicy-

lic acid in *Penicillium patulum*. The only exception is represented by pamamycin, which, in addition to its antibiotic activity, is also an inducer of differentiation.

A metabolic function has been suggested for the ionophore antibiotics, such as polyethers and sideramines. Under particular conditions these may substitute for the natural carriers of ions such as potassium and iron. However, an experimental demonstration of this function is lacking, except in the case of a strain of *Streptomyces griseus*, which produces a macrotetrolide, carrier of potassium. Actually, it is more likely that "siderophore" antibiotics have no function in ion transport, but that, on the contrary, they sequester ions, competing with natural siderophores.

9.3.3. Inhibition of Other Organisms

According to this hypothesis, the evolutionary advantage conferred by antibiotics would be that of simply being antibiotics, i.e., inhibitors of microorganisms that compete with the producer in a given ecological niche. This concept is widely accepted for other secondary metabolites, e.g., alkaloids of plants, which are toxic for animals that would use them as nutriment or for parasites, but it is not universally accepted as far as microorganisms are concerned. In particular, it has been maintained that antibiotics are produced effectively only under laboratory conditions and not in nature.

The following considerations can be put forward in support of this hypothesis.

1. Today, it is sufficiently documented that antibiotics are produced in the natural environment. Furthermore, it appears unreasonable to assume that the expression of numerous and accurately regulated genes such as those that code for the antibiotic biosynthesis takes place under artificial conditions only.
2. The majority of antibiotics are produced by microorganisms that dwell in the soil, in a competitive situation. Organisms such as archebacteria, which live in particular environments (high temperature and acidity, high salt concentration), do not produce antibiotics; gram-negative enterobacteria do not generally produce antibiotics but may produce bacteriocins, proteins that are specifically active against a small number of other enterobacteria.
3. Antibiotics that are active against different genera of microorganisms are frequently produced by the same strain. Many antibacterial antibiotics are produced together with antifungal and antiprotozoal polyenes.

4. When, in particular, the life cycle of streptomycetes is taken into consideration, one can note that formation of aerial mycelium takes place at the expense of vegetative mycelium, which lyses yielding the necessary nutrients. It is reasonable to assume that there is a mechanism that prevents the utilization of these nutrients by other microorganisms. This observation also explains the temporal association between antibiotic production and the beginning of differentiation.

In conclusion, it is reasonable to believe that the function of antibiotics, i.e., of the secondary metabolites possessing a high and specific antimicrobial activity, is of an ecological nature: antibiotic production would give a competitive advantage vis-à-vis the other microorganisms sharing the same environment. Other secondary metabolites may have different functions: mycotoxins, for instance, although provided with a low antibiotic activity, appear to play an important role in the relation between molds and plants; other metabolites such as protease inhibitors, or some compounds recently isolated from fungi, may have a regulation function or transfer biochemical information between cells.

Chapter 10

Further Reading

Chapter 1. The Antibiotics: An Overview

Books

Bérdy, J., Aslalos, A., Bostian, M., and McNitt, K., 1981, *Handbook of Antibiotic Compounds*, Vols. I–IX, CRC Press, Boca Raton, Fla.

Bétina, V., 1983, *The Chemistry and Biology of Antibiotics*, Elsevier, Amsterdam.

Bycroft, B. W. (ed.), 1988, *Dictionary of Antibiotics and Related Substances*, Chapman & Hall, London.

Demain, A. L., and Solomon, N. A. (eds.), 1983, *Antibiotics Containing the β-Lactam Structure*, Vols. 1 and 2, Springer-Verlag, Berlin.

Hlavka, J. J., and Boothe, J. H., 1985, *The Tetracyclines*, Springer-Verlag, Berlin.

Kleinkauf, H., and von Düren, H. (eds.), 1990, *Biochemistry of Peptide Antibiotics*, de Gruyter, Berlin.

Laskin, A. I., and Lechevalier, H. A. (eds.), 1988, *Handbook of Microbiology*, 2nd ed., Vol. IX, Part A, CRC Press, Boca Raton, Fla.

Morin, R. B., and Gorman, M. (eds.), 1982, *Chemistry and Biology of β-Lactam Antibiotics*, Vols. 1 and 2, Academic Press, New York.

Omura, S. (ed.), 1984, *Macrolide Antibiotics*, Academic Press, New York.

Pape, H., and Rehm, H. J. (eds.), 1986, *Biotechnology*, Vol. 4: *Microbial Products II*, VCH Verlag, Weinheim.

Umezawa, H., and Hooper, I. R. (eds.), 1982, *Aminoglycoside Antibiotics*, Springer-Verlag, Berlin.

Periodicals

Antimicrobial Agents and Chemotherapy, American Society for Microbiology, Washington, D.C.
The Journal of Antibiotics, Japan Antibiotics Research Association, Tokyo.
The Journal of Antimicrobial Chemotherapy, British Society for Antimicrobial Chemotherapy, London.

Chapter 2. The Activity of Antibiotics

Balows, A., Hausler, W. J., Hermann, K. L., Isemberg, H. D., and Shadomy, H. J., 1991, *Manual of Clinical Microbiology*, 5th ed., American Society for Microbiology, Washington, D.C.
Collins, C. H., and Lyne, P. M., 1984, *Microbiological Methods*, 5th ed., Butterworths, London.
Hewitt, W., and Vincent, S., 1989, *Theory and Application of Microbiological Assay*, Academic Press, New York.
Lorian, V. (ed.), 1986, *Antibiotics in Laboratory Medicine*, 2nd ed., Williams & Wilkins, Baltimore.
Phillips, I. (chairman), 1991, A guide to sensitivity testing, *J. Antimicrob. Chemother.* **27** (Suppl. D):1.
Reeves, D. S., Phillips, I., Williams, J. D., and Wise, R., 1978, *Laboratory Methods in Antimicrobial Chemotherapy*, Churchill, Livingstone, Edinburgh.

Chapter 3. The Mechanism of Action of Antibiotics

General

Franklin, T. J., and Snow, G. A., 1989, *Biochemistry of Antimicrobial Action*, 4th ed., Chapman & Hall, London.
Gale, E. F., Cundliffe, E., Reynolds, P. E., Richmond, M. H., and Waring, M. J., 1981, *The Molecular Basis of Antibiotic Action*, Wiley, New York.
Greenwood, D., and O'Grady, F. (eds.), 1985, *The Scientific Basis of Antimicrobial Chemotherapy*, Cambridge University Press, London.
Hahn, F. E. (ed.), 1979, *Antibiotics V*, Vols. 1 and 2, Springer-Verlag, Berlin.
Hahn, F. E. (ed.), 1983, *Antibiotics VI: Modes and Mechanisms of Microbial Growth Inhibitors*, Springer-Verlag, Berlin.
Kerridge, D., 1986, Mode of action of clinically important antifungal drugs, *Adv. Microb. Physiol.* **27**:1.
Russel, A. D., and Chopra, I., 1990, *Understanding Antibacterial Action and Resistance*, Ellis Horwood, Chichester.

Specific Mechanisms

Cannon, M., 1990, Agents which interact with ribosomal RNA and interfere with its function, in *Comprehensive Medicinal Chemistry*, Vol. 2 (P. G. Sammes, ed.), pp. 814–838, Pergamon Press, Oxford.

Drlica, K., 1984, Biology of bacterial deoxyribonucleic acid topoisomerases, *Microbiol. Rev.* **48**:273.
Hertzberg, R. P., 1990, Agents interfering with DNA enzymes, in *Comprehensive Medicinal Chemistry*, Vol. 2 (P. G. Sammes, ed.), pp. 753–791, Pergamon Press, Oxford.
Hobbes, J. B., 1990, Purine and pyrimidine targets, in *Comprehensive Medicinal Chemistry*, Vol. 2 (P. G. Sammes, ed.), pp. 299–332, Pergamon Press, Oxford.
Nikaido, H., and Vaara, M., 1985, Molecular basis of bacterial outer membrane permeability, *Microbiol. Rev.* **45**:1.
Sensi, P., and Lancini, G. C., 1990, Inhibitors of transcribing enzymes: Rifamycins and related agents, in *Comprehensive Medicinal Chemistry*, Vol. 2 (P. G. Sammes, ed.), pp. 793–811, Pergamon Press, Oxford.
Spratt, B. G., 1983, Penicillin-binding proteins and the future of β-lactam antibiotics, *J. Gen. Microbiol.* **129**:1247.
Tanaka, N., 1982, Mechanism of action of aminoglycosides antibiotics, in *Aminoglycosides Antibiotics* (H. Humezawa and I. R. Hooper, eds.), pp. 221–266, Springer-Verlag, Berlin.
Ward, J. B., 1990, Cell wall structure and function, in *Comprehensive Medicinal Chemistry*, Vol. 2 (P. G. Sammes, ed.), pp. 553–607, Pergamon Press, Oxford.

Chapter 4. Resistance of Microorganisms to Antibiotics

Bryan, L. E. (ed.), 1984, *Antimicrobial Drug Resistance*, Academic Press, New York.
Bryan, L. E. (ed.), 1989, *Microbial Resistance to Drugs*, Springer-Verlag, Berlin.
Bush, K., 1989, Characterization of β-lactamases, *Antimicrob. Agents Chemother.* **33**:259.
Bush, K., 1989, Classification of β-lactamases 1, 2a, 2b, and 2b′, *Antimicrob. Agents Chemother.* **33**:264.
Bush, K., 1989, Classification of β-lactamases 2c, 2d, 2e, 3, and 4, *Antimicrob. Agents Chemother.* **33**:271.
Foster, T. J., 1983, Plasmid determined resistance to antimicrobial drugs and toxic metal ions in bacteria, *Microbiol. Rev.*, **47**:361.
Franklin, T. J., and Snow, G. A., 1989, *Biochemistry of Antimicrobial Action*, 4th ed., Chapman & Hall, London.
Leclerq, R., and Courvalin, P., 1989, Bacterial resistance to macrolide, lincosamide, streptogramin antibiotics by target modification, *Antimicrob. Agents Chemother.* **35**:735.
Lyon, B. R., and Scurray, R., 1987, Antimicrobial resistance in *S. aureus*: Genetic basis, *Microbiol. Rev.*, **51**:88.
Quesnel, L. B., 1990, Resistance and tolerance to antimicrobial drugs, in *Comprehensive Medicinal Chemistry*, Vol. 2 (P. G. Sammes, ed.), pp. 89–122, Pergamon Press, Oxford.
Russel, A. D., and Chopra, I, 1990, *Understanding Antibacterial Action and Resistance*, Ellis Horwood, Chichester.

Chapter 5. Activity of Antibiotics in Relation to Their Structure

General

Betina, V., 1983, *The Chemistry and Biology of Antibiotics*, Elsevier, Amsterdam.
Kucers, A., and McBennet, N., 1987, *The Use of Antibiotics*, 4th ed., Heinemann Medical Books, London.

Antibiotic Families

Brown, A. G., Pearson, M. J., and Southgate, R., 1990, Other β-lactam agents, in *Comprehensive Medicinal Chemistry*, Vol. 2 (C. Hansch, P. G. Sammes, and J. B. Taylor, eds.), pp. 655–702, Pergamon Press, Oxford.

Demain, A. L., and Solomon N. A. (eds.), 1983, *Antibiotics Containing the β-Lactam Structure*, Vols. 1 and 2, Springer-Verlag, Berlin.

Hlavka, J. J., and Boothe, J. H., 1985, *The Tetracyclines*, Springer-Verlag, Berlin.

Kirst, H., and Sides, G. D., 1989, New directions for macrolide antibiotics: Structural modifications and *in vitro* activities, *Antimicrob. Agents Chemother.* **33**:1413.

Lancini, G. C., and Cavalleri, B., 1990, Glycopeptide antibiotics of the vancomycin group, in *Biochemistry of Peptide Antibiotics* (H. Kleinkauf and H. von Dören, eds.), pp. 159–178, de Gruyter, Berlin.

Morin, R. B., and Gorman, M. (eds.), 1982, *Chemistry and Biology of β-Lactam Antibiotics*, Vols. 1 and 2, Academic Press, New York.

Neu, H. C., 1986, β-Lactam antibiotics: Structural relationships affecting *in vitro* activity and pharmacologic properties, *Rev. Infect. Dis.* **8**:S237.

Newall, C. E., and Hallam, P. D., 1990, β-Lactam antibiotics: Penicillins and cephalosporins, in *Comprehensive Medicinal Chemistry*, Vol. 2 (C. Hansch, P. G. Sammes, and J. B. Taylor, eds.), pp. 609–653, Pergamon Press, Oxford.

Omura, S. (ed.), 1984, *Macrolide Antibiotics*, Academic Press, New York.

Sensi, P., and Lancini, G. C., 1990, Inhibitors of transcribing enzymes: Rifamycins and related agents, in *Comprehensive Medicinal Chemistry*, Vol. 2 (C. Hansch, P. G. Sammes, and J. B. Taylor, eds.), pp. 793–811, Pergamon Press, Oxford.

Tsukagoshi, S., Takeuci, T., and Umezawa, H., 1986, Antitumor substances in *Biotechnology*, Vol. 4 (H. Pape and H. J. Rehm, eds.), pp. 509–530, VCH Verlag, Weinheim.

Umezawa, H., and Hooper, I. R. (eds.), 1982, *Aminoglycoside Antibiotics*, Springer-Verlag, Berlin.

von Dören, H., 1990, Compilation of peptide structures. A biogenetic approach, in *Biochemistry of Peptide Antibiotics* (H. Kleinkauf and H. von Dören, eds.), pp. 411–507, de Gruyter, Berlin.

Ward, J. B., 1990, Cell wall structure and functions, in *Comprehensive Medicinal Chemistry*, Vol. 2 (C. Hansch, P. G. Sammes, and J. B. Taylor, eds.), pp. 553–607, Pergamon Press, Oxford.

Weiss, R. B., Sarosy, G., Clagett-Carr, K., Russo, M., and Leyland-Jones, B., 1986, Anthracycline analogues: The past, present, and future, *Cancer Chemother. Pharmacol.* **18**:185.

Chapter 6. Biosynthesis and Genetics of Antibiotic Production

General

Baltz, R. H., Hegeman, G. D., and Skatrud, P. L. (eds.), 1993, *Industrial Microorganisms: Basic and Applied Molecular Genetics*, American Society for Microbiology, Washington, D.C.

Corcoran, J. W. (ed.), 1981, *Antibiotics IV: Biosynthesis*, Springer-Verlag, Berlin.

Floss, H. G., and Beale, J. M., 1989, Biosynthetic studies on antibiotics, *Angew. Chem. Int. Ed. Engl.* **28**:146.

Hershberger, C. L., Queener, S. W., and Hegeman, G. (eds.), 1989, *Genetics and Molecular Biology of Industrial Microorganisms*, American Society for Microbiology, Washington, D.C.

Horinouchi, S., and Beppu, T., 1992, Autoregulatory factors and communication in actinomycetes, *Annu. Rev. Microbiol.* **46**:377.

Lancini, G. C., and Lorenzetti, R., 1993, *Biotechnology of Antibiotics and Other Microbial Metabolites*, Plenum Press, New York.

Pape, H., and Rehm, H. J. (eds.), 1986, *Biotechnology*, Vol. 4: *Microbial Products II*, VCH Verlag, Weinheim.

Vandamme, E. J. (ed.), 1984, *Biotechnology of Industrial Antibiotics*, Dekker, New York.

Vining, L. C. (ed.), 1983, *Biochemistry and Genetic Regulation of Commercially Important Antibiotics*, Addison–Wesley, Reading, Mass.

Specific Pathways

Aharonowitz, Y., Cohen, G., and Martin, J. F., 1992, Penicillin and cephalosporin biosynthetic genes: Structure, organization, regulation, and evolution, *Annu. Rev. Microbiol.* **46**:461.

Donadio, S., Staver, M. J., McAlpine, J. B., Swanson, S. J., and Katz, L., 1991, Modular organization of genes required for complex polyketide biosynthesis, *Science* **252**:675.

Doull, J., Ahmed, Z., Stuttard, C., and Vining, L. C., 1985, Isolation and characterization of *Streptomyces venezuelae* mutants blocked in chloramphenicol biosynthesis, *J. Gen. Microbiol.* **131**:97.

Ebersole, R. C., Godfredsen, W. O., Vangedal, S., and Caspi, E., 1973, Mechanism of oxidative cyclization of squalene. Evidence for cyclization of squalene from either end of squalene molecule in the *in vivo* biosynthesis of fusidic acid by *Fusidium coccineum*, *J. Am. Chem. Soc.* **95**:8133.

Elson, S. W., Baggaley, K. H., Davison, M., Fulstone, M., Nicholson, N. H., Risbridger, G. D., and Tyler, J. W., 1993, The identification of three new biosynthetic intermediates and one further biosynthetic enzyme in the clavulanic acid pathway, *J. Chem, Soc., Chem. Commun.* **1993**:1212.

Harris, C. M., Roberson, J. S., and Harris, T. M., 1976, Biosynthesis of griseofulvin, *J. Am. Chem. Soc.* **98**:5380.

Isono, K., 1988, Nucleoside antibiotics: Structure, antibiotic activity and biosynthesis, *J. Antibiot.* **41**:1711.

Jung, G., 1991, Lantibiotics—Ribosomally synthesized biologically active polypeptides containing sulfide bridges and α-β-didehydroamino acids, *Angew. Chem. Int. Ed. Engl.* **30**:1051.

Kakinuma, K., Ogawa, Y., Sakasi, T., Seto, H., and Otake, N., 1989, Mechanism and stereochemistry of the biosynthesis of 2-deoxystreptamine and neosamine C, *J. Antibiot.* **42**:926.

Katz, L., and Donadio, S., 1993, Polyketide synthesis: Prospects for hybrid antibiotics, *Annu. Rev. Microbiol.* **47**:875.

Kleinkauf, H., and von Döhren, H., 1987, Biosynthesis of peptide antibiotics, *Annu. Rev. Microbiol.* **41**:259.

Kuo, M. S., Yurek, D. A., Coats, J. H., Chung, S. T., and Li, G. P., 1992, Isolation and identification of 3-propylidene-Δ^1-pyrroline-5-carboxylic acid, a biosynthetic precursor of lincomycin, *J. Antibiot.* **45**:1773.

Lacalle, R. A., Tercero, J. A., and Jiménez, A., 1992, Cloning of the complete biosynthetic gene cluster for an aminonucleoside antibiotic, puromycin, and its regulated expression in heterologous hosts, *EMBO J.* **11**:785.

Lancini, G. C., 1986, Ansamycins, in *Biotechnology*, Vol. 4: *Microbial Products II* (H. Pape and H.-J. Rehm, eds.), pp. 431–463, VCH Verlag, Weinheim.

Lancini, G. C., 1989, Fermentation and biosynthesis of glycopeptide antibiotics, *Prog. Ind. Microbiol.* **27**:283.

Martin, J. F., 1984, Biosynthesis, regulation and genetics of polyene macrolide antibiotics, in *Macrolide Antibiotics* (S. Omura, ed.), pp. 405–424, Academic Press, New York.

Okuda, T., and Ito, Y., 1982, Biosynthesis and mutasynthesis of aminoglycoside antibiotics, in *Aminoglycoside Antibiotics* (H. Umezawa and I. R. Hooper, eds.), pp. 111–203, Springer-Verlag, Berlin.

Omura, S., and Tanaka, Y., 1984, Biochemistry, regulation and genetics of macrolide production, in *Macrolide Antibiotics* (S. Omura, ed.), pp. 199–259, Academic Press, New York.

Perlman, D., Otani, S., Perlman, K. L., and Walker, J. E., 1973, 3-Hydroxy-4-methylkynurenine as an intermediate in actinomycin biosynthesis, *J. Antibiot.* **26**:289.

Vater, J., 1990, Gramicidin S synthetase, in *Biochemistry of Peptide Antibiotics* (H. Kleinkauf and H. von Döhren, eds.), pp. 33–55, de Gruyter, Berlin.

Walker, J. B., 1975, Pathways of biosynthesis of guanetidated inositol moieties of streptomycin and bluensomycin, *Methods Enzymol.* **43**:429.

Chapter 7. The Search for and Development of New Antibiotics

Search for New Antibiotics

Bu'Lock, J. D., Nisbett, L. J., and Wisteinley, D. J. (eds.), 1982, *Bioactive Microbial Products: Source and Discovery*, Academic Press, New York.

Cross, T., 1982, Actinomycetes: A continuing source of new metabolites, *Dev. Ind. Microbiol.* **23**:1.

Iwai, Y., and Omura, S., 1982, Culture conditions for screening of new antibiotics, *J. Antibiot.*, **35**:123.

Lancini, G. C., and Lorenzetti, R., 1993, *Biotechnology of Antibiotics and Other Bioactive Microbial Metabolites*, pp. 73–93, Plenum Press, New York.

Lee, G. P., 1989, Isolation of actinomycetes for antibiotic screening, *Chin. J. Antibiot.*, **14**:452.

Sutcliffe, J. A., and Georgopapadakou, N. H. (eds.), 1992, *Emerging Targets in Antibacterial and Antifungal Chemotherapy*, Chapman & Hall, London.

Development from the Laboratory to the Clinic

Cleeland, R., and Grunberg, E., 1986, Laboratory evaluation of new antibiotics *in vitro* and in experimental animal infections, in *Antibiotics in Laboratory Medicine*, 2nd ed. (V. Lorian, ed.), pp. 825–876, Williams & Wilkins, Baltimore.

Gootz, T. D., 1990, Discovery and development of new antimicrobial agents, *Clin. Microb. Rev.* **1990**:13.

Zak, O., and O'Reilly, T., 1991, Animal models in evaluation of antimicrobial agents, *Antimicrob. Agents Chemother.* **35**:1527.

Development from the Laboratory to the Manufacturing Process

Bader, F. G., 1986, Physiology and fermentation development, in *The Bacteria*, Vol. IX: *Antibiotic Producing Streptomyces* (S. W. Queener and L. E. Day, eds.), pp. 281–321, Academic Press, New York.

Chater, K. F., 1990, The improving prospect for yield increase by genetic engineering in antibiotic producing streptomyces, *Biotechnology* **8**:115.

Lancini, G. C., and Lorenzetti, R., 1993, *Biotechnology of Antibiotics and Other Bioactive Microbial Metabolites*, pp. 175–190, Plenum Press, New York.

Nisbett, L. J., and Winstanley, D. J. (eds.), 1983, *Bioactive Microbial Products 2: Development and Production*, Academic Press, New York.

Normansell, I. D., 1986, Isolation of streptomyces mutants improved for antibiotic production, in *The Bacteria*, Vol. IX: *Antibiotic Producing Streptomyces* (S. W. Queener and L. E. Day, eds.), pp. 95–118, Academic Press, New York.

Chapter 8. The Use of Antibiotics

Clinical Uses

Greenwood, D. (ed.), 1989, *Antimicrobial Chemotherapy*, 2nd ed., Oxford University Press, London.

Kucers, A., McBennet, N., and Kemp, R. J., 1987, *The Use of Antibiotics*, 4th ed., Heinemann, London.

Kuemmerle, H. P. (ed.), 1983, *Clinical Chemotherapy*, Vols. 1–3, Thieme–Stratton, New York.

Pratt, W. B., and Fekety, R., 1986, *The Antimicrobial Drugs*, Oxford University Press, London.

Uses in Animal Husbandry and in Agriculture

Berdy, J., 1986, Further antibiotics with practical application, in *Biotechnology*, Vol. 4 (H. Pape and H. J. Rehm, eds.), pp. 487–505, VCH Verlag, Weinheim.

Braude, R., 1981, Antibiotics as feed additives for livestock, in *The Future of Antibiotherapy and Antibiotic Research* (L. Ninet, P. E. Bost, D. H. Bounchaud, and J. Florent, eds.), pp. 169–182, Academic Press, New York.

Misato, T., Ko, K., and Yamaguchi, I., 1988, Use of antibiotics in agriculture, *Adv. Appl. Microbiol.* **21**:53.

Ruckebusch, R., 1981, Antimicrobial drugs in veterinary medicine, in *The Future of Antibiotherapy and Antibiotic Research* (L. Ninet, P. E. Bost, D. H. Bounchaud, and J. Florent, eds.), pp. 141–167, Academic Press, New York.

Vandamme, E. J. (ed.), 1984, *Biotechnology of Industrial Antibiotics*, Dekker, New York.

Chapter 9. Antibiotics and Producer Organisms

Bennet, J., and Bentley, R., 1989, What is a name? Microbial secondary metabolism, *Adv. Appl. Microbiol.* **34**:1.

Cundliffe, E., 1989, How antibiotic-producing organisms avoid suicide, *Annu. Rev. Microbiol.* **43**:207.

Davies, J., 1990, What are antibiotics? Archaic functions for modern activities, *Mol. Microbiol.*, **4**:1227.

Demain, A. L., 1989, Functions of secondary metabolites, in *Genetics and Molecular Biology of Industrial Microorganisms* (C. L. Herschberger, S. W. Queener, and G. Hegeman, eds.), pp. 1–11, American Society for Microbiology, Washington, D.C.

Laskin, A. I., and Lechevalier, H. A. (eds.), 1988, *Handbook of Microbiology*, 2nd ed., Vol. 9, Part A, CRC Press, Boca Raton, Fla.

Martin, J. F., and Liras, P., 1989, Organization and expression of genes involved in the biosynthesis of antibiotics and other secondary metabolites, *Annu. Rev. Microbiol.* **43**:173.

Stone, M. J., and Williams, D. H., 1992, On the evolution of functional secondary metabolites, *Mol. Microbiol.*, **6**:29.

Vining, L. C., 1990, Functions of secondary metabolites, *Annu. Rev. Microbiol.* **44**:395.

Index

MIX
Papier aus verantwortungsvollen Quellen
Paper from responsible sources
FSC® C105338

If you have any concerns about our products,
you can contact us on
ProductSafety@springernature.com

In case Publisher is established outside the EU,
the EU authorized representative is:
Springer Nature Customer Service Center GmbH
Europaplatz 3, 69115 Heidelberg, Germany

Printed by Libri Plureos GmbH
in Hamburg, Germany